Contemporary Theories of Knowledge

Second Edition

JOHN L. POLLOCK
AND
JOSEPH CRUZ

ROWMAN & LITTLEFIELD PUBLISHERS, INC.
Lanham • Boulder • New York • Oxford

ROWMAN & LITTLEFIELD PUBLISHERS, INC.

Published in the United States of America
by Rowman & Littlefield Publishers, Inc.
4720 Boston Way, Lanham, Maryland 20706

12 Hid's Copse Road
Cumnor Hill, Oxford OX2 9JJ, England

British Library Cataloguing in Publication Information Available

Library of Congress Cataloging-in-Publication Data

Pollock, John L.
    Contemporary theories of knowledge / John L. Pollock and Joseph
Cruz.
        p. cm. — (Studies in epistemology and cognitive theory)
    Includes bibliographical references and index.
    ISBN 0-8476-8936-0 (cloth : alk. paper)
    ISBN 0-8476-8937-9 (pbk : alk. paper)
    1. Knowledge, Theory of. 2. Philosophy, Modern—20th century. I,
Cruz, Joseph, 1969- II. Title. III. Series: Studies in epistemology
and cognitive theory (Unnumbered)
    BD 161 .P7245   1999
    121—dc21   99-12525
                                                          CIP

Printed in the United States of America

♾™ The paper used in this publication meets the minimum requirements of American
National Standard for Information Sciences—Permanence of Paper for Printed Library
Materials, ANSI/NISO Z39.48–1992.

For Cynthia, Mark, and Melissa

CONTENTS

Preface

This book is intended to play two roles. On the one hand, it is a textbook. It is intended as an introduction to the theory of knowledge for readers with some intellectual sophistication but without an extensive knowledge of philosophy. We do not think that this goal is incompatible with a commitment to scholarship, but we have attempted to be sensitive to the fact that some technical issues are tangential to the main strands of the book. We have occasionally relegated issues and sources that will be of interest primarily to advanced readers to footnotes and hope that our liberal attitude toward footnotes will be useful to those pursuing further research in epistemology.

On the other hand, this book is an attempt to say what is true in epistemology, and in this latter guise it is aimed as much at the professional philosopher as at the student.

The book endeavors to play both roles by taking its principal task to be that of mapping out the logical geography of epistemology in a way that enables the reader to see how the issues and theories fit together. A taxonomy of epistemological theories is constructed, and then the different kinds of theories are discussed in terms of their place in the taxonomy. We have done our best to present each general variety of epistemological theory in the best possible light. Then we have tried to raise only those objections to the theories that reflect very general features of them. We have tried to avoid raising objections that might be met by tinkering with details. In this way we have been led to reject all of the more familiar kinds of epistemological theories (foundations theories, coherence theories, probabilist theories, and reliabilist theories). That exhausts most of the logical geography of epistemology and leaves us with only a small verdant landscape to explore—the region of what we call "nondoxastic internalist theories". These are theories that insist that the justifiability of a belief is a function exclusively of the internal states of the believer, but also insist that we must include more than the believer's beliefs among those internal states. In particular, the believer's perceptual states and memory states can be relevant to what she is justified in believing even when she has no beliefs about what perceptual states and memory states she is in.

We begin with a discussion of the problems of knowledge by reviewing skepticism and by rehearsing some of the particular conundrums that have motivated epistemology. We then turn to standard issues in the theory of justification. In chapter two, we discuss foundationalism. Chapters three and four concern coherentism and externalism, respectively. These first four chapters follow a course that will be familiar from other treatments of topics in epistemology. Starting in chapter five, however, we offer a set of discussions that seem to us to be neglected in epistemology but sorely needed. We attempt to describe what epistemic normativity is, and how norms are able to regulate our epistemic conduct. Too often,

epistemologists seem satisfied to limp along with a conception of normativity that is inherited from metaethics or, worse, to avoid the question
entirely. We hope that this chapter will be of interest to a wide philosophical audience, as normativity is a crucial component of philosophical
theorizing.

In chapter six we reflect on the questions of what epistemic rationality
is and what the methodology of epistemology should be. Epistemologists
sometimes conduct their research without critically evaluating how they
go about using philosophical intuitions to create a theory of knowledge
or justification. It might reasonably be asked, "why should we think
that using philosophical intuitions will reveal anything about knowledge?"
We want to face that challenge directly. In addition, the discussion of
methodology is an appropriate time to introduce one of the central topics
of epistemology in the last decade, namely the status of epistemological
naturalism. Our view is naturalistic, but our naturalism is quite different
from the views that are commonly associated with that label.

One of our main concerns in this book is to integrate epistemology
into a naturalistic view of human beings as a kind of biological information
processor. We have tried to do this by exploring how we might build an
intelligent machine that is capable of interacting with its environment
and surviving in a hostile world. Very general constraints having to do
with limited computational powers lead to a machine many of whose
features reproduce initially surprising aspects of human epistemology.
The epistemology of such a machine will almost automatically be a variety
of the kind of naturalistic nondoxastic internalism that we have described.
In chapter seven, we explore the details of the epistemic norms that arise
out of the insights gained by taking this design stance in epistemology.
The seventh chapter is presented in a style that may be unfamiliar to
philosophical readers, as we do not shy away from taking a stand on the
specific implementational details of rational norms. We offer this discussion because we view working out the details of epistemic norms in the
context of AI to be a powerful tool for revealing our own rationality. We
maintain that philosophers should not be averse to employing a range of
intellectual tools in their understanding of human knowledge. While we
grant that it is possible to skip the seventh chapter without misunderstanding our position, we hope that students and professionals alike will
appreciate the need for providing a logically precise specification of our
epistemic norms.

To a certain extent the core of this book grew out of two journal
articles by Pollock: "Epistemic Norms" (*Synthese*) and "My Brother, the
Machine" (*Nous*). The former provides the positive theory of the book,
and some material from the latter provides, in a sense, the theoretical
underpinning for the positive theory. Jointly, they comprise chapter
five. The rest of the book consists of a discussion of competing theories
and was written as more or less an introduction to chapter five. In the
end, the discussion of competing theories became as important as the
positive theory because of the light the discussion throws on the general

structure of epistemology and epistemological problems. The second edition extends the theory developed in the first edition by incorporating some of the results of the OSCAR project. Chapters six and seven incorporate material from several recent journal articles and book chapters, and rely upon results defended in Pollock's *Cognitive Carpentry*. We wish to thank the publishers of the following articles for allowing us to reprint parts of these articles: "Procedural epistemology—at the interface of Philosophy and AI" (*Blackwell Guide to Epistemology*), "Procedural epistemology" (*The Digital Phoenix: How Computers Are Changing Philosophy*), "Reasoning about change and persistence: a solution to the frame problem" (*Nous*), "The theory of nomic probability"(*Synthese*), and "Justification and defeat" (*Artificial Intelligence*).

We have many debts to acknowledge in this second edition. Some of these debts are shared, as they were incurred during a time when we were both at the University of Arizona. Still, it is easiest at this point for us to express ourselves separately.

The first edition of this book profited from philosophical discussion of its topics with many of my colleagues and students. Those who stand out most prominently in my mind are Keith Lehrer, Alvin Goldman, Steven Schiffer, Stewart Cohen, John Carroll, George Smith, Bob Audi, and Hilary Kornblith. Others too numerous to mention have helped me in clarifying my thoughts on various aspects of the material and have helped me avoid errors I would otherwise have fallen into.

The further advances reported in this second edition are largely a result of work performed in the OSCAR project, which has been heavily supported by the University of Arizona and the National Science Foundation (grant no. IRI-9634106). I particularly want to thank Merrill Garrett who, in his position as the Director of Cognitive Science at the University of Arizona, has been unstinting in his support.

— John Pollock
Tucson, Arizona

My thinking about issues in epistemology has been influenced by wonderful philosophers and cognitive scientists both at the University of Arizona and elsewhere. I am especially indebted to Alvin Goldman, Keith Lehrer, and Rob Cummins, three of my mentors who reveal as much in their agreements as they do in their disagreements. I have also profited tremendously from discussions and correspondence with Karen Wynn, Jack Lyons, Joel Pust, Linda Radzik, Tim Bayne, Neil Stillings, and Jonathan Vogel.

Melissa Barry has been a constant source of support and keen philosophical insight. Our conversations on naturalism and normativity have shaped my thinking in what I hope she regards as a positive way. Her wider philosophical influence on me cannot be repaid.

Finally, I owe my greatest thanks to my co-author, John Pollock. I have been deeply affected over the years by his intellect, his wit, and his outdoor aesthetic. Most of all, I have been touched by his respect for me, and by his enduring friendship.

During the writing of this book I received support from an NEH faculty development grant administered by Hampshire College.

— Joe Cruz
Northampton, Massachusetts

1
THE PROBLEMS
OF KNOWLEDGE

1. Cognition

What sets human beings apart from other animals is their capacity for sophisticated thought. Only human beings are capable of the kinds of cognition required to build an airplane or a microwave oven, to write *Hamlet* or compose a symphony, to propound the Theory of Relativity or discover DNA. We have voluminous knowledge of the world, and most of that knowledge concerns matters other animals cannot even conceive of. How is it that we are able to engage in such sophisticated thought and arrive at the capacious knowledge that we use to direct both our everyday activities and our momentous achievements like flying to the moon or curing cancer? That is the subject matter of epistemology—the theory of knowledge.

What we want is to understand rational thought, from the routine to the sublime. We want to know how it is possible for us to accomplish the epistemic tasks upon whose results we base our lives and which other creatures find so impossible. We want to understand human beings as cognizers.

Cognitive psychology investigates certain aspects of human cognition through the methods of science. But our interest here is in specifically *rational* cognition. Psychologists study human thought when it goes wrong as well as when it goes right, but we want to know *what it is* for it to go right. What is it that makes human beings rational, and thereby makes our enormous intellectual achievements possible? The psychologist, in studying both rational and irrational thought, presupposes a pre-theoretic understanding of rationality, but does little to illuminate it. While we may marvel at the cognitive psychologist's use of sophisticated experiments to determine how human beings think, what it means to think rationally is a philosophical challenge. What we want is a general theory of rationality—what is it to be a rational cognizer, and how does being rational make it possible for us to acquire the wide variety of world knowledge that we take almost for granted? This is a philosophical question with a long and rich history.

Rational cognition includes more than the pursuit of knowledge. Knowledge has a purpose. It is to help us get around in the world. We use our knowledge to guide us in deciding how to act, and rational cognition includes the cognitive processes involved in action decisions. This book, however, will focus on the purely intellectual aspects of cognition that are involved in the pursuit of knowledge.

One of the remarkable conclusions of contemporary epistemology is that the rational thought responsible for our great intellectual achievements is not different in kind from the rational thought involved in routine epistemic procedures, like determining the color of an object seen in broad daylight, remembering your mother's name, discovering that most objects fall to the ground when unsupported, or summing 12 and 25. If we can understand how rational thought enables us to solve these routine epistemic problems, we can understand the discovery of DNA as the result of stringing together a large number of routine problems. What is extraordinary about human thought is already present in our ability to solve routine epistemic problems. Thus we begin by focusing on the mundane, in hopes that it will lead us to the sublime.

2. Skeptical Problems

Rather than ask, "How is it possible to discover DNA, or find a cure for cancer?", the epistemologist has traditionally begun by asking, "How is knowledge possible at all?" The philosophically inexperienced reader might find this puzzling, thinking to herself, "I know how to tell what color something is and how to remember my mother's name. I don't care about that. I want to know how to find a cure for cancer." Indeed, we all know how to perform simple epistemic tasks, but there is an important difference between knowing how to do it and knowing how it is done. We all know how to pick up a cup of coffee without spilling it, but imagine trying to give a precise description of how to do this sufficient to enable an engineer to build an industrial robot to accomplish the same task. In fact, engineers spent years trying to solve this very problem. Similarly, although we know how to perform simple epistemic tasks, it is extremely difficult to explain how we do and why what we do yields knowledge. Historically, philosophers have often motivated the study of simple epistemic tasks with the help of *skeptical arguments*. These are initially compelling arguments that seem to show that even simple epistemic tasks are impossible. Consider the following tale:

> *It all began that cold Wednesday night. I was sitting alone in my office watching the rain come down on the deserted streets outside, when the phone rang. It was Harry's wife, and she sounded terrified. They had been having a late supper alone in their apartment when suddenly the front door came crashing in and six hooded men burst into the room. The men were armed and they made Harry and Anne lay face down on the floor while they went through Harry's pockets. When they found his driver's license one of them carefully scrutinized Harry's face, comparing it with the official photograph and then muttered, "It's him all right." The leader of the intruders produced a hypodermic needle and injected Harry with something that made him lose consciousness almost immediately. For some reason they only tied and gagged Anne. Two of the*

men left the room and returned with a stretcher and white coats. They put Harry on the stretcher, donned the white coats, and trundled him out of the apartment, leaving Anne lying on the floor. She managed to squirm to the window in time to see them put Harry in an ambulance and drive away.

By the time she called me, Anne was coming apart at the seams. It had taken her several hours to get out of her bonds, and then she called the police. To her consternation, instead of uniformed officers, two plain clothed officials arrived and, without even looking over the scene, they proceeded to tell her that there was nothing they could do and if she knew what was good for her she would keep her mouth shut. If she raised a fuss they would put out the word that she was a psycho and she would never see her husband again.

Not knowing what else to do, Anne called me. She had had the presence of mind to note down the number of the ambulance, and I had no great difficulty tracing it to a private clinic at the outskirts of town. When I arrived at the clinic I was surprised to find it locked up like a fortress. There were guards at the gate and it was surrounded by a massive wall. My commando training stood me in good stead as I negotiated the 20 foot wall, avoided the barbed wire, and silenced the guard dogs on the other side. The ground floor windows were all barred, but I managed to wriggle up a drainpipe and get in through a second-story window that someone had left ajar. I found myself in a laboratory. Hearing muffled sounds next door I peeked through the keyhole and saw what appeared to be a complete operating room and a surgical team laboring over Harry. He was covered with a sheet from the neck down and they seemed to be connecting tubes and wires to him. I stifled a gasp when I realized that they had removed the top of Harry's skull. To my horror one of the surgeons reached into the open top of Harry's head and eased his brain out, placing it in a stainless steel bowl. The tubes and wires I had noted earlier were connected to the now disembodied brain. The surgeons carried the bloody mass carefully to some kind of tank and lowered it in. My first thought was that I had stumbled on a covey of futuristic Satanists who got their kicks from vivisection. My second thought was that Harry was an insurance agent. Maybe this was their way of getting even for the increases in their malpractice insurance rates. If they did this every Wednesday night, their rates were no higher than they should be!

My speculations were interrupted when the lights suddenly came on in my darkened hidey hole and I found myself looking up at the scariest group of medical men I had ever seen. They manhandled me into the next room and strapped me down on an operating table. I thought, "Uh, oh, I'm in for it now!" The doctors huddled at the other end of the room, but I couldn't turn my head far enough to see what they were doing. They were mumbling among themselves, probably deciding my fate. A door opened and I heard a woman's voice. The deferential manner assumed by the medical malpractitioners made it obvious who was boss. I strained to see this mysterious woman but she hovered just out of my view. Then, to my astonishment, she walked up and stood over me and I realized it was my secretary, Margot. I began to wish I had given her that Christmas bonus after all.

It was Margot, but it was a different Margot than I had ever seen. She was wallowing in the heady wine of authority as she bent over me. "Well Mike, you thought you were so smart, tracking Harry here to the clinic," she said. Even now she had the sexiest voice I have ever heard, but I wasn't really thinking about that. She went on, "It was all a trick just to get you here. You saw what happened to Harry. He's not really dead, you know. These gentlemen are the premier neuroscientists in the world today. They have developed a surgical procedure whereby they remove the brain from the body but keep it alive in a vat of nutrient. The Food and Drug Administration wouldn't approve the procedure, but we'll show them. You see all the wires going to Harry's brain? They connect him up with a powerful computer. The computer monitors the output of his motor cortex and provides input to the sensory cortex in such a way that everything appears perfectly normal to Harry. It produces a fictitious mental life that merges perfectly into his past life so that he is unaware that anything has happened to him. He thinks he is shaving right now and getting ready to go to the office and stick it to another neurosurgeon. But actually, he's just a brain in a vat."

"Once we have our procedure perfected we're going after the head of the Food and Drug Administration, but we needed some experimental subjects first. Harry was easy. In order to really test our computer program we need someone who leads a more interesting and varied life —someone like you!" I was starting to squirm. The surgeons had drawn around me and were looking on with malevolent gleams in their eyes. The biggest brute, a man with a pockmarked face and one beady eye staring out from under his stringy black hair, was fondling a razor sharp scalpel in his still-bloody hands and looking like he could barely restrain his excitement. But Margot gazed down at me and murmured in that incredible voice, "I'll bet you think we're going to operate on you and remove your brain just like we removed Harry's, don't you? But you have nothing to worry about. We're not going to remove your brain. We already did—three months ago!"

With that they let me go. I found my way back to my office in a daze. For some reason, I haven't told anybody about this. I can't make up my mind. I am racked by the suspicion that I am really a brain in a vat and all this I see around me is just a figment of the computer. After all, how could I tell? If the computer program really works, no matter what I do, everything will seem normal. Maybe nothing I see is real. It's driving me crazy. I've considered checking into that clinic voluntarily and asking them to remove my brain just so that I can be sure. Frankly, I don't know if even that would put my worries to rest.

Mike is luckier than most brain-in-a-vat victims. He at least has a clue to his precarious situation—Margot told him he is a brain in a vat. Of course, it could all be contrived. Perhaps he is not a brain in a vat after all. There is no way he can be sure. Meditating about this case, it may occur to you that you might be a brain in a vat, too. If you are, there is no way you could ever find out. Nor, it seems, is there any way you can be sure you are not a brain in a vat, because everything would seem just the same to you in either case. But if you cannot be sure you are not a brain in a vat, how can you trust the evidence of your senses? You have no way of knowing that they are not figments of a computer. It seems that you cannot *really* know anything about the world around you. It could all be an illusion. You cannot rule out the possibility that you are a brain in a vat, and without being able to rule out that possibility, knowledge of the material world is impossible.

This is a typical example of a skeptical problem. Mike's plight involves fanciful technology and thinking about it is entertaining, but conundrums of this sort have a serious philosophical point. Skeptical problems seem to show that we cannot have the kinds of knowledge we are convinced we have, including the most mundane sorts of knowledge that we take for granted on a daily basis. If skeptical problems challenge our most basic kind of knowledge, then they appear also to easily and completely undermine the sophisticated knowledge that is distinctive of human beings. Such problems have played a central role in epistemology. It is tempting to become caught up in the task of refuting the skeptic, and at one time epistemologists took that to be their principal goal. Descartes was concerned with finding beliefs that he could not reasonably doubt and to which he could appeal in justifying all the rest of his beliefs, and Hume was nonplussed by his inability to answer his own skeptical dilemma about induction. In the *Critique of Pure Reason*, Kant wrote:

> It still remains a scandal to philosophy ... that the existence of things outside of us ... must be accepted merely on *faith*, and that, if anyone thinks good to doubt their existence, we are unable to counter his doubts by any satisfactory proof.[1]

But contemporary epistemology tends to take a different attitude toward skepticism. If we consider a variety of skepticism that confines itself to some limited class of beliefs, it *might* be possible to answer the skeptic by showing that those beliefs can be securely defended by appeal to other beliefs not among those deemed problematic. But for any very general kind of skepticism, that is impossible in principle. Every argument must proceed from some premises, and if the skeptic calls all relevant premises into doubt at the same time then there is no way to reason with him. The whole enterprise of refuting the skeptic is ill-founded, because he will not allow us anything with which to work.

The proper treatment of skeptical arguments requires looking at them in a different light. We come to philosophy with a large stock of beliefs. Initially, we regard them all as knowledge, but then we discover that they conflict. They cannot all be true because some are inconsistent with others. One instance of this general phenomenon is represented by skeptical arguments. Starting from premises in which we are initially confident,

---

1. Kant (1958), p. 34. This passage is quoted by G. E. Moore (1959), p. 126.

the skeptical argument leads us to the conclusion that we cannot possibly
have certain kinds of knowledge. But we are also initially confident that
we do have such knowledge. Thus our original confidently held beliefs
form an inconsistent set. We cannot reasonably continue to hold them
all.

Upon discovering that our system of beliefs is inconsistent, the initial
reaction might be that we should throw them all away and start over
again. Descartes pursued this strategy at the beginning of his *Meditations
on First Philosophy*. But that will not solve the problem. The skeptic is
not just questioning our beliefs. He is also questioning the cognitive
processes by which we arrive at our beliefs, and if we start all over again
we will still be employing the same cognitive processes. We cannot
dispense with *both* the beliefs and the cognitive processes, because then
we would have nothing with which to begin again. As Otto Neurath
(1932) put it in an often-quoted passage, "We are like sailors who must
rebuild their ship upon the open sea."[2] We must start with the beliefs
and cognitive processes we have and repair them "from within" as best
we can. The legitimacy of beginning with what we already have was
urged by G. E. Moore in a famous passage:

> I can prove now, for instance, that two human hands exist. How? By
> holding up my two hands, and saying, as I make a certain gesture with
> the right hand, "Here is one hand," and adding, as I make a certain
> gesture with the left, "and here is another." ... But now I am perfectly
> well aware that, in spite of all that I have said, many philosophers will
> still feel that I have not given any satisfactory proof of the point in
> question. ... If I had proved the propositions which I used as *premisses* in
> my two proofs, then they would perhaps admit that I had proved the
> existence of external things. ... They want a proof of what I assert *now*
> when I hold up my hands and say "Here's one hand and here's another."
> ... They think that, if I cannot give such extra proofs, then the proofs
> that I have given are not conclusive proofs at all. ... Such a view, though
> it has been very common among philosophers, can, I think, be shown to
> be wrong. ... I can know things which I cannot prove; and among things
> which I certainly did know, ... were the premisses of my two proofs. I
> should say, therefore, that those, if any, who are dissatisfied with these
> proofs merely on the ground that I did not know their premisses, have
> no good reason for their dissatisfaction. (1959, 144ff)

Even though Moore's complete hostility to skepticism has difficulties of
its own, there is something to his point. If we reflect upon our beliefs,
we will find that we are more confident of some than of others. It is
reasonable to place more reliance on those beliefs in which we have
greater confidence, and when beliefs come in conflict we decide which
to reject by considering which we are least certain of. If we have to reject
something, it is reasonable to reject those beliefs we regard as most

---

2. "Wie Schiffer sind wir, die ihr Schiff auf offener See umbauen müssen." This
passage has been immortalized by Quine (1960) who refers to it repeatedly.

doubtful.[3] Now consider how these observations apply to skeptical arguments. An argument begins from premises and draws a conclusion:

$$P_1$$
$$P_2$$
$$\vdots$$
$$P_n$$

Therefore, Q.

Presented with an argument whose premises we believe, the natural
reaction is to accept the conclusion, even if the conclusion is the denial of
something else we initially believe. But that is not always the reasonable
response to an argument. In the above argument, Q is a deductive
consequence of $P_1,...,P_n$, but all that really shows is that we cannot reasonably continue to believe all of $P_1,...,P_n$ and $\sim Q$. The validity of the argument
does not establish which of these beliefs should be rejected, because we
can convert the argument into an equally valid argument for the denial
of any one of the premises. For instance, the following is also a valid
argument (where $\sim Q$ is short for 'It is false that Q'):

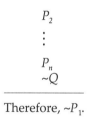

$$P_2$$
$$\vdots$$
$$P_n$$
$$\sim Q$$

Therefore, $\sim P_1$.

Faced with a skeptical argument, we believe all of the premises $P_1,...,P_n$,
but we also believe $\sim Q$ (the denial of the conclusion, the conclusion
being that we do not have the knowledge described). The argument
establishes that we must reject one of these beliefs, but it does not tell us
which we should reject. To determine that, we must reflect upon how
certain we are of each of these beliefs and reject the one of which we are
least certain. In typical skeptical arguments, we invariably find that we
are more certain of the knowledge seemingly denied us than we are of
some of the premises. Thus it is not reasonable to adopt the skeptical
conclusion that we do not have that knowledge. The rational stance is
instead to deny one or more of the premises. In other words, a typical
skeptical argument is best viewed as a *reductio ad absurdum* of its premises,
rather than as a proof of its conclusion.[4]

---

3. This is what John Rawls (1971) calls "the method of reflective equilibrium".
4. There is no logical necessity that this should be the case. It is conceivable that

To illustrate, consider inductive reasoning. When we reason inductively, we draw general conclusions from the observation of a finite number of instances. For example, having observed many swans, and noting that they were all white, we may infer that all swans are white. This is typical of the kind of reasoning that is performed in science when we test scientific theories by testing whether particular instances of them are true. Virtually all of our general beliefs about the world are held on the basis of induction. But an obvious feature of inductive reasoning is that the truth of the premises does not logically guarantee that the general conclusion is true. For example, in concluding that all swans are white we would be mistaken, because it turns out that there are black swans in Australia.

In the *Treatise on Human Nature*, Hume used the above observations to propound a skeptical argument against induction. He reasoned as follows:

1. The premises of an inductive argument do not logically entail the conclusion.
2. If the premises of an argument do not logically entail the conclusion, then it is not reasonable to believe the conclusion on the basis of the premises.

3. Therefore, inductive reasoning is illegitimate—one cannot acquire knowledge of general truths by reasoning inductively.

The conclusion of this argument is something that we initially disbelieve. That is, a little reflection convinces us that we know many general truths about the world on the basis of induction. However, the premises of the argument can also seem convincing. The first premise is illustrated by the swan example. And there was a time in the not too distant past when virtually all philosophers took the second premise to be equally obvious. If the premises of an argument can be true without the conclusion being true, then how (they asked) could the premises give us any reason for believing the conclusion?

This is another typical skeptical argument. The premises seem initially compelling, but the conclusion that we cannot have inductive knowledge seems absurd. The strategy suggested above is to treat this as a *reductio ad absurdum* argument. Although the premises seem initially compelling, they seem less certain than the conclusion seems certainly false, so we should take the argument as an argument to the effect that one of the

there should be a skeptical argument whose premises we believe more firmly than we believe that we have the putative knowledge the argument denies us. The claim we are making here is a contingent one about those skeptical arguments that have actually been advanced in philosophy.

THE PROBLEMS OF KNOWLEDGE

premises is false. The epistemological problem then becomes that of deciding which premise is false, and why.

A few philosophers have tried to respond to Hume's argument by denying the first premise. Bertrand Russell (1912) and C. I. Lewis (1956) both suggested that inductive reasoning is based upon an additional premise that turns the inductive argument into a deductive argument. The premise they sought was called "the uniformity of nature". This premise was sometimes formulated rather vaguely as the principle that the future will be like the past. The intention was for this principle of the uniformity of nature to be sufficiently strong that, when conjoined with the inductive evidence, it would provide us with premises logically entailing the inductive generalization. So, for example, because we have in the past observed only white swans, we should expect that swans observed in the future will also be white.

The most obvious difficulty for trying to resolve Hume's skeptical dilemma in this way is that of trying to give a precise formulation of an appropriate principle of the uniformity of nature. It is not sufficient to say that the future will be like the past, because obviously the future isn't *always* like the past. Things change.

But an ultimately more compelling objection is that no principle of the uniformity of nature could do the job required of it. It is supposed to turn inductive reasoning into deductively valid reasoning. Thus in the case of the swans, our actual argument will be as follows:

> Swan #1 is white.
> Swan #2 is white.
>
> :
>
> Nature is uniform.
> _____
> Therefore, all swans are white.

The difficulty is that the conclusion of this argument is false. If it is deductively valid, it must have a false premise. But all the premises other than that of the uniformity of nature are true, so the premise of the uniformity of nature must be false. As such, it becomes useless to us in reasoning about the world. It is not reasonable to accept conclusions drawn by reasoning from a premise known to be false. Thus if this were the correct form of inductive reasoning, all inductive reasoning would be based upon a false premise, and we would be led to Hume's skeptical conclusion all over again, by a different route.

If we are to avoid Hume's conclusion, we cannot do it by denying the first premise of his argument. Thus it must be the second premise that is false. Contrary to what Hume supposed, it must be possible for the premises of an argument to support a conclusion without logically entailing it. In other words, the premises of the argument can provide a reason for believing the conclusion without the reason being logically conclusive. Such reasons are *defeasible* in the sense that, while they can justify us in believing their conclusions, that justification can be "defeated" by acquiring further relevant information. In the case of induction, if we observe that all the A's in our sample are B's, this may provisionally justify us in believing that all A's are B's. But if we subsequently encounter another A and note that it is not a B, that is sufficient to defeat the original justification and it makes it totally unreasonable to continue believing that all A's are B's. Thus the skeptical argument about induction points to the existence of defeasible reasons in epistemology. In our opinion, one of the most important advances in epistemology in the last half of the twentieth century was the recognition of defeasible reasons. It is now generally acknowledged that most of our reasoning proceeds defeasibly rather than deductively. Induction is one example of this, but other examples are equally obvious. For example, our beliefs about our surroundings are based upon perception. When we perceive the world around us, it looks various ways to us, and in the absence of conflicting information we conclude that it is the way it looks. But obviously, our senses can mislead us. The world does not *have* to be the way it looks. So our reasoning is defeasible rather than deductive. Defeasible reasoning will be discussed at some length in the next chapter.

This discussion of Hume's argument illustrates the importance of skeptical arguments for epistemology. What makes skeptical arguments important is not that their conclusions might be true. They are important for what they show about knowledge rather than because they make us doubt that we have knowledge. The task of the contemporary epistemologist is to *understand* knowledge. For this she need not refute the skeptic—we already know that the skeptic is wrong. Nevertheless, important conclusions about the nature of knowledge and epistemic justification can be gleaned from the investigation of skeptical arguments. This is because such an argument constitutes a *reductio ad absurdum* of its premises, and its premises consist of things we initially believe about knowledge and justification. Thus in deciding which of those premises is wrong we are learning something new about knowledge and correcting mistaken beliefs with which we begin. In short, the task of the epistemologist is not to show *that* the skeptic is wrong but to explain *why* he is wrong. The difference between these endeavors is that in the latter we can take it as a premise that we have various kinds of knowledge (i.e., we can assume ~Q) and see what that requires. For example, we might ask, "Given that we have perceptual knowledge, what must the relationship be between our perceptual beliefs and our sensory experience?" The fact that we do have perceptual knowledge will impose important constraints on that relationship and can lead us to significant conclusions about epistemic justification. This reasoning has the form, "We do have such-and-such knowledge; we could not have that knowledge if so-and-so were the case; therefore, so-and-so is not the case." This kind of reasoning is very common in contemporary epistemology. Note that such reasoning results from contraposing the premises and conclusion of a skeptical argument.

3. Knowledge and Justification

We have been discussing the role of skeptical arguments in epistemology and have already made some progress in thinking about reasoning and belief. Although skepticism has in the past played a defining role in epistemology, addressing skepticism is still merely one project that epistemology faces. We can now step back a bit from this discussion in order to get a broader sense of contemporary theories of knowledge. In light of the myriad other developments in epistemology in the philosophical literature, perhaps the only genuinely uncontroversial thing that can be said of epistemology as whole is that it is an attempt to make sense of the possibility and limits of human intellectual achievement. Traditionally, achievements of the intellect are associated with knowledge. And since epistemology is "the theory of knowledge", it would seem most naturally to have knowledge as its principal focus. But that is not entirely accurate. The theory of knowledge is an attempt to answer the question, "How do you know?", but this is a question about *how* one knows, and not about knowing per se. In asking how a person knows something we are typically asking for her grounds for believing it. We want to know what *justifies* her in holding her belief. Thus epistemology has traditionally focused on epistemic justification more than on knowledge. Epistemology might better be called "doxastology", which means the study of beliefs. The epistemic agent is viewed in epistemology as capable of representing the world through his or her beliefs and other mental states, where these are taken as thoughts that can be more or less rational to maintain. The philosophical status of mental states is a central issue in the philosophy of mind, though some philosophers have attempted to tackle both epistemological problems and philosophy of mind problems simultaneously.[5] Later in this book, we will offer a longer discussion of our views on mental states like beliefs.

A justified belief is one that it is "epistemically permissible" to hold. Epistemic justification is a normative notion. It pertains to what you *should* or *should not* believe. But it is a uniquely epistemic normative notion. Epistemic permissibility must be distinguished from both moral and prudential permissibility. For example, because beliefs can have important consequences for the believer, it may be prudent to hold beliefs for which you have inadequate evidence. For instance, it is popularly alleged that lobsters do not feel pain when they are dunked alive into boiling water. It is extremely doubtful that anyone has good reason to believe that, but it may be prudentially rational to hold that belief because otherwise one would deprive oneself of the gustatory delight of eating boiled lobsters. Conversely, it may be imprudent to hold beliefs for which you have unimpeachable evidence. Consider Helen, who has

---

5. See the theories of Fred Dretske (1981), Gilbert Harman (1973), and William Lycan (1988).

overwhelming evidence that her father is Jack the Ripper. It may be that if she admitted this to herself it would be psychologically crushing. In such cases people sometimes do not believe what the evidence overwhelmingly supports. That is prudentially reasonable but epistemically unreasonable. Thus epistemic reasonableness is not the same thing as prudential reasonableness. Epistemic reasonableness is also distinct from moral reasonableness. It is unclear whether moral considerations can be meaningfully applied to beliefs. If not, then epistemic justification is obviously distinct from moral permissibility. If belief does fall within the purview of morality then presumably a belief can be made morally impermissible, for example, if one were to promise someone never to think ill of him. But clearly the moral permissibility of such a belief is totally unrelated to its epistemic permissibility. This is not to say that it is inappropriate to ask whether certain beliefs are morally permissible. Our point is that this is an enterprise separate from epistemology. Epistemic justification is normative, but it must be distinguished from other familiar normative concepts.

Epistemic justification governs what you should or should not believe. Rules describing the circumstances under which it is epistemically permissible to hold beliefs are called *epistemic norms*. An important task of epistemology is that of describing the epistemic norms governing various kinds of belief. For instance, philosophers have sought accounts of the circumstances under which it is epistemically permissible to believe, on the basis of sense perception, that there are physical objects of different sorts standing in various spatial relations to the perceiver. In part, epistemologists have tried to elicit the nature of the epistemic norms governing this kind of knowledge by looking at skeptical arguments purporting to show that perceptual knowledge is impossible. We know, contrary to the skeptic, that perceptual knowledge *is* possible, and that allows us to draw conclusions about the epistemic norms governing perceptual knowledge. This will be a recurring theme throughout the book.

If the central question of epistemology concerns the justification of belief rather than knowledge, why is the discipline called "epistemology"? The explanation lies in the fact that there appear to be important connections between knowledge and justification. We have already noted that the question "How do you know?" can generally be construed as meaning "What justifies you in believing?", but we can reasonably ask why it can be construed in that way. To answer this question, epistemologists have spent a great deal of time laboring over the connections between knowledge and justification. One way of thinking about the connection is to view knowledge as an achievement that can be further understood as having good reasons for a belief. If one's reasons for a belief are good, then the belief is justified. So, it has been generally acknowledged that epistemic justification is a necessary condition for knowledge.[6] Of course,

---

6. Not quite all philosophers acknowledge this. There are a variety of strategies

putting the project this way immediately demands an account of   .
makes something a reason and what makes some reasons good, and that
is just what a theory of justification aspires to.

Consensus is rare in philosophy, but from the early part of this century
until 1963 it was almost universally agreed that knowledge was the same
thing as justified true belief. That is, a person knows something, *P*, if
and only if (1) she believes it, (2) it is true, and (3) her belief is justified.
But in 1963, Edmund Gettier published his seminal paper "Is justified
true belief knowledge?" in which he showed to everyone's astonishment
that this identification is incorrect.[7] He did this by presenting counterexamples. In one of his examples we consider Smith, who believes falsely
but with good reason that Jones owns a Ford. Smith has no idea where
Brown is, but he arbitrarily picks Barcelona and infers from the putative
fact that Jones owns a Ford that either Jones owns a Ford or Brown is in
Barcelona. It happens by chance that Brown is in Barcelona, so this
disjunction is true. Furthermore, as Smith has good reason to believe
that Jones owns a Ford, he is justified in believing this disjunction. But
as his evidence does not pertain to the true disjunct of the disjunction,
we would not regard Smith as *knowing* that it is true that either Jones
owns a Ford or Brown is in Barcelona.

The general reaction to Gettier's examples has been to concede that a
fourth condition must be added to the analysis of "*S* knows that *P*". The
search for this fourth condition has become known as the *Gettier problem*.
The Gettier problem is a seductive sort of problem, much as the problem
of skepticism is seductive. When they first encountered it, most epistemologists were convinced that it must have a simple solution. Simple
conditions were found that handled the original Gettier counterexamples,
but new counterexamples emerged almost immediately. More and more
complicated counterexamples were followed by more and more complicated fourth conditions.[8] At the present time, the Gettier problem has
become mired in complexity and few philosophers now expect it to have
a simple solution. Nevertheless, having gotten hooked on the problem
epistemologists are loathe to let it go, so it remains a frequent topic in
contemporary epistemology.

The Gettier problem has fundamentally altered the character of con-

---

for detaching knowledge from justification. For accounts that attempt to analyze
knowledge by a strategy where a belief must co-vary with the truth in a way that does
not involve justification see Peter Unger (1967), Joseph Margolis (1973), Alvin Goldman
(1976), David Armstrong (1973), and Robert Nozick (1981). Colin Radford (1966) inspired
a debate on this topic by offering putative cases of knowledge where the epistemic
agent is not justified because she does not believe that which she knows. For replies,
see David Annis (1969), David Armstrong (1969), and Keith Lehrer (1968).

7. Remarkably, a counterexample to the traditional analysis can also be found in
Bertrand Russell (1912), p. 132, but that went overlooked.

8. For a thorough discussion of the Gettier literature through the early 1980's, see
Robert Shope (1983).

temporary epistemology. Many epistemologists now regard the Gettier problem as a central problem of epistemology since it poses a clear barrier to analyzing knowledge. It is our conviction, however, that this represents an important and lamentable shift in the focus of epistemology. Historically, the central topic of epistemology was epistemic justification rather than knowledge.[9] Philosophers were more interested in *how* we know than *what it is* to know. Of course, this is in part because they thought that the latter question had an easy answer. The Gettier problem shows that the answer is not easy. The analysis of "*S* knows that *P*" is a fascinating problem, but it should be regarded as a side issue rather than as the central problem of epistemology. What the Gettier problem really shows is what a perverse concept knowledge is. One can do everything with complete epistemic propriety, and be right, and yet lack knowledge because of some accident about the way the world is. Why do we employ such a concept? For obvious reasons, we will often be interested in whether someone is right in some belief he or she holds, and we may be interested in whether the person is justified in his or her belief because that may be indicative of how reliable they are in their knowledge claims. But if we know that they are both right and justified, why should we have any further interest in whether they know? Clearly we often do, and why we do is an interesting puzzle, but it is not the sort of question that should be regarded as the founding question for an entire discipline.[10] Epistemology was traditionally concerned with how rational cognition works in forming and updating our beliefs about the world, and that should once again be recognized as the central problem of epistemology and one of the main questions of philosophy. For this reason the Gettier problem is not discussed in this book. Its solution is not an essential part of the construction of an epistemological theory. On the contrary, we take an epistemological theory to be a theory about how it is possible to acquire various kinds of knowledge, and this is most basically a theory about epistemic justification.

It is useful to distinguish two potentially different concepts of epistemic justification. We have taken the fundamental problem of epistemology to be that of deciding what to believe. Epistemic justification, as we use the term, is concerned with this problem. Considerations of epistemic justification guide us in determining what to believe. We might call this the "procedural" sense of "justification". Correlatively, epistemic norms are norms prescribing how to form beliefs. It is common in contemporary epistemology to find philosophers explaining instead that what they mean by "justification" is, roughly, "what turns true belief into knowledge". That, of course, is not very clear. In any event, it must be emphasized that the topic of this book is procedural justification, not

---

9. This is persuasively documented by Mark Kaplan (1985).
10. Ruth Millikan (1984) and Stephen Stich (1992) offer similar concerns about how contemporary epistemology treats knowledge.

"what is required for knowledge". This will be important at various stages of the argument where it is urged that particular theories could not play a procedural role.

4. Areas of Knowledge

We know many kinds of things, and there appear to be important differences between the ways we know them. We can subdivide knowledge into different "areas", according to these epistemological differences. Knowledge based directly upon sense perception, or "perceptual knowledge", comprises one area. Knowledge possessed by virtue of remembering previously acquired knowledge comprises another. Inductive generalizations comprise a third. Knowledge of other minds, a priori knowledge, and moral knowledge comprise other areas. Knowledge in different areas will share some common features but will also exhibit important differences.

4.1 Perceptual Knowledge

The *problem of perception* is that of explaining how perceptual knowledge is possible. We all agree that sense perception can lead to justified beliefs about the world around us. But the details remain obscure. The skeptical argument with which this chapter began can be regarded as an assault on the possibility of perceptual knowledge. It seems that our perceptual experience could be precisely what it is without the world being at all what it appears to be (we might be brains in vats!). How then is it possible to acquire knowledge of the material world by relying upon sense perception?

The focus of the present book is of "meta-epistemology". That is, it is more concerned with describing and contrasting *kinds* of epistemological theories than it is with addressing specific epistemological problems. The broad categories of epistemological theories that will be discussed will be enumerated in the next section. But one way of contrasting theories is by comparing what they have to say about specific epistemological problems, and the sample problem to which we will return repeatedly is the problem of perception. More has been written in epistemology about perceptual knowledge than about any other kind of knowledge. This is partly because the psychological facts are clearer. Specifically, we know that such knowledge is acquired in response to the activation of our sense organs, the most important of which is vision. This enables us to formulate the problem of perception as that of explaining how we can acquire justified beliefs about the external world on the basis of the output of our sense organs. This seemingly unremarkable formulation contrasts sharply with the formulation of epistemological problems concerning some other areas of knowledge.

4.2 A Priori Knowledge

In contrast to perceptual knowledge, even the very basic psychological facts about other areas of knowledge tend to be obscure. It is clear that sense perception is the source of perceptual knowledge, but for some areas of knowledge the source is quite mysterious. A priori knowledge comprises one of the most problematic areas. *A priori knowledge* is usually defined as "what is known independently of experience", or perhaps as "what is known on the basis of reason alone". But it must be acknowledged that these are not very helpful definitions and they should not be taken too seriously. Rather, we recognize that there is a certain class of knowledge that seems importantly different from other kinds of knowledge and we give it a label—"a priori knowledge". The class is characterized by its stereotypes. These include most prominently knowledge of mathematical and logical truths. It is very difficult to say in even a superficial way what is involved psychologically in the acquisition of a priori knowledge. For instance, consider mathematical knowledge. We know that mathematical proof is an important factor in mathematical knowledge. The nature of mathematical proof is itself fraught with difficulty, but an even more obscure aspect of mathematical knowledge arises from the observation that any substantive proof (i.e., any proof of something other than a principle of logic) must start from premises already established. Where do the basic premises of mathematics come from? A once-popular view was that they are arbitrary axioms laid down by convention and that they "implicitly define" mathematical concepts.[11] Such a "conventionalist" view was attractive because it seemed to reduce a priori knowledge to something much easier to understand. But conventionalism lost its plausibility, partly because of Gödel's theorem. A rough formulation of Gödel's theorem is that given any set of axioms for mathematics, there are theorems we can prove in "real mathematics" that cannot be deduced from those axioms. This seems to show that we have more mathematical knowledge than we could have if conventionalism were true.[12] Today conventionalism has few supporters.

Ordinarily, the downfall of one theory heralds the apparent success of another. Theories are rarely overturned except in the face of seemingly better theories. But in the case of a priori knowledge, no better theory has appeared on the horizon. Other than conventionalism, the only kind of theory that has occurred to people is what might be called "a priori intuitionism".[13] According to this theory, basic a priori truths are "self-evident". We have the power to "intuit" that they are true. This putative faculty of a priori intuition has been described variously by different

---

11. See for example A. J. Ayer (1946).

12. For a detailed discussion of conventionalism and its relationship to Gödel's theorem, see Pollock (1974), chapter ten.

13. This is to be distinguished from the philosophy of mathematics known as "mathematical intuitionism" and defended by such people as Brouwer and Heyting. There is no close connection between these two kinds of intuitionism.

philosophers. Bertrand Russell described it as the power to directly intuit relations between universals.[14] Other authors have tried to describe it in a more ontologically neutral way.[15] But notice that the claim that we have such a faculty at all is really a psychological claim. Furthermore, although it is one that psychologists have not directly addressed, it must be regarded as being at least somewhat suspect. If there is any such faculty of a priori intuition, it tends to elude introspection. We would not claim at this point that there definitely is no such faculty, or that a priori intuitionism is a false theory, but it must be acknowledged that the psychological facts surrounding a priori knowledge are obscure. This makes it difficult to either formulate or evaluate philosophical theories of a priori knowledge.[16]

4.3 Moral Knowledge

A priori knowledge is not the only area in which the psychological facts are obscure. Moral knowledge is at least as problematic. There is not even a consensus that moral knowledge exists. Although some moral philosophers are convinced that there is such a thing as moral knowledge, at least as many are adamant that there is not. The latter philosophers maintain that moral language plays a unique role that does not involve expressing truths. It has been urged, for example, that moral language expresses sentiments, or approval and disapproval, or some other kind of psychological attitude distinct from belief.[17] In any such "nonobjective" moral theory there is no such thing as an epistemological problem of moral knowledge, because if there are no moral truths then there can be no moral knowledge.

Suppose we set aside nonobjective views and assume that there is such a thing as moral knowledge. Then we are faced with explaining how we can acquire that knowledge. One possible view is analogous to a priori intuitionism. According to *ethical intuitionism*, we have a faculty of moral intuition that makes some moral truths self-evident, and then other moral truths can be defended on the basis of the self-evident ones.[18] But there is no general agreement that we have a psychological faculty of moral intuition. Ethical intuitionism is not popular in contemporary philosophy. There are alternative theories, but none of them are very popular either. The psychological foundations of putative moral knowledge are in disarray. But without a better understanding of the psychological facts surrounding moral reasoning it is hard to get a philosophical theory of moral knowledge off the ground.

---

14. One representative of such a view was Bertrand Russell (1912).
15. This course was taken in Pollock (1974).
16. An excellent discussion of a priori knowledge can be found in Laurence Bonjour (1998). Jerrold Katz (1998) also discusses a priori knowledge with a special emphasis on the epistemology of mathematics.
17. See for example A. J. Ayer (1946) and Charles Stevenson (1944).
18. The most important proponents of ethical intuitionism are H. A. Prichard (1950), Sir David Ross (1930), and G. E. Moore (1903).

18 CHAPTER ONE

4.4 Knowledge of Other Minds

Solipsism is the view that there is only a single mind in the universe and that it is one's own. Solipsism is a lonely doctrine and if someone firmly maintained it she might very well be driven mad. Showing that solipsism is false has generated its own skeptical puzzle, namely the *problem of other minds*. The problem of other minds is to give a satisfying account of how we know that there are other minds in the world. Perception gives us information about the behavior of creatures that appear to be very much like ourselves. And sometimes perception gives us information about the brain and nervous system of human beings and other animals. But this is not the same as showing that there are other minds.

In a way, the status of our knowledge of other minds is the inverse of the status of a priori knowledge and moral knowledge. The psychological facts regarding our willingness to claim knowledge of other minds are relatively clear, at least in broad outline. Psychologically we are prone to attribute a mind to anything that behaves in a characteristic way. Certain patterns of motion, episodes of sustained adaptive behavior or even possession of expressive eyes seem to propel us to the claim that something has a mind.[19] This temptation is so seductive that we even attribute minds to things that we know do not have minds such as the forces of nature or ordinary desktop computers.[20] Some psychologists propose that we reason that other people and entities have minds on the basis of an analogy with our own case. While this proposal is in need of further empirical evidence, even if it were shown to be true the epistemic propriety of reasoning by analogy in this case would need to be addressed.[21] Thus, knowledge of other minds is a distinct area of knowledge.

4.5 Memory

Much of what we know, we know by remembering. This has suggested to some epistemologists that memory is a source of knowledge. According to this view, remembering involves a psychological state—what we might call "apparent memory"—that plays a role in memory analogous to the role sense perception plays in perceptual knowledge.[22] Other philosophers have disputed this claim, insisting that memory introduces no new source of knowledge. Instead, they maintain, memory is just the exercise of previously acquired knowledge, and the source of remembered knowledge is whatever the source of the knowledge was when it was first acquired.[23]

---

19. See Simon Baron-Cohen (1994).
20. Daniel Dennett emphasizes our psychological propensity to attribute other minds in his (1987) and (1995).
21. Wittgenstein's *Philosophical Investigations* is associated with skepticism about reasoning by analogy in the case of other minds. See especially (1953), ¶293. Pollock (1974) defends such analogical reasoning.
22. This view was defended in Pollock (1974), chapter seven.
23. This view was defended by Norman Malcolm (1963, pp. 229-230), and

The debate here is in part over what occurs psychologically in remembering. Is there an introspectible state of apparent memory that distinguishes remembering something from simply believing it? Philosophers do not agree, and psychologists have done little to resolve the issue. But even if the psychological facts were resolved in favor of apparent memory, the philosophical question would remain whether apparent memory somehow licenses belief in what one presently remembers, or whether instead the justification of current memories is the same as the justification of those same beliefs when they were originally acquired.

Memory comprises an area of knowledge in the same sense that perceptual knowledge and a priori knowledge do, but memory also has a more pervasive significance for meta-epistemology. One of the main ways in which epistemological theories differ from one another is in their account of reasoning and its relationship to epistemic justification. Memory plays a fundamental role in reasoning. When we reason in accordance with any even slightly complicated argument, we do not hold the entire argument in mind at the same time. We attend to each step individually and rely upon memory to tell us that we got to that step in some reasonable way. A correct epistemological account of memory must make this legitimate. Thus the nature of memory knowledge will play a pivotal role in the formulation of alternative epistemological theories. For this reason, memory will be discussed at some length in the next chapter.

4.6 Induction

Knowledge of inductive generalizations comprises a kind of knowledge importantly different from the varieties of knowledge described so far. Induction is distinguished not by its source but by its method. The simplest kind of induction is *enumerative induction*, wherein we examine a sample of objects of some kind, A, observe that all the A's in the sample have another property, B, and infer on that basis that all A's are B's. A related kind of induction is *statistical induction*, where instead of observing that all the A's in our sample are B's we observe that some proportion m/n of them are B's and then infer that the probability of an arbitrary A being a B is approximately m/n. The way we reason in enumerative and statistical induction is fairly clear in bold outline, but the fine details have been remarkably resistant to accurate description.

Induction has exercised philosophers because of two different kinds of worries. The *traditional problem of induction* is Hume's problem, discussed above. Hume took his problem to be that of answering the skeptic —a task at which he confessed defeat. In light of our earlier discussion of the role of skeptical arguments in contemporary epistemology, we can dismiss the traditional problem in the form given it by Hume. There is neither a need nor the possibility of proving the skeptic wrong. You can

Robert Squires (1969).

never *prove* the skeptic wrong because he does not leave you with enough ammunition to undertake such a task. But there is no reason why we should have to prove the skeptic wrong. We already know that he is wrong. One of the things we are certain about right from the beginning is that we can acquire knowledge of general truths on the basis of induction.

Even if we dismiss skepticism regarding induction, we can resurrect Hume's problem in a new guise. While there is no need to justify induction in the sense of proving that inductive reasoning is epistemologically legitimate, it may still be possible to justify induction in another sense. The question arises whether inductive reasoning is a fundamental and irreducible component of our framework of reasons and reasoning.[24] If it is then no further justification can be demanded, but if it is not then it may be possible to base inductive reasoning on simpler and more basic kinds of reasoning. There is reason to think that the latter alternative may be the correct one. Principles of induction seem simple until we try to formulate them precisely; then they become extremely complicated and it is never clear whether we have got them quite right. We will illustrate some of the difficulties below. What is to be noted here is that if correct principles of induction are really that complicated, and we have that much trouble telling whether we have formulated them correctly, then it is unlikely that they are formulations of basic epistemic norms we follow directly in our reasoning. Instead, it seems likely that they reflect the application of simpler epistemic norms to cases having enough internal complexity to render the application logically and mathematically convoluted.[25]

Leaving aside the problem of justifying induction, much philosophical labor has also gone into what Nelson Goodman (1955) dubbed *the new riddle of induction*. This is the problem of giving an accurate formulation of principles of induction. Goodman's main interest was in just one aspect of this problem, but it is a multi-faceted problem. The aspect that concerned Goodman was the *problem of projectibility*. This is best illustrated by thinking first about deductive inferences. Rules of deductive inference apply equally to all propositions and properties. For example, we can infer Q from $(P \& Q)$ regardless of what P and Q are. The traditional view of induction took it to be like deduction in applying equally to all properties, but Goodman startled the world of philosophy by showing that there are restrictions on the use of inductive reasoning. Goodman's examples were highly contrived, but quite simple examples are available. For instance, having observed a sample of ravens, you might note that all the ravens you have observed have been observed. Obviously, that gives you no reason to believe that all ravens have been observed. Other

---

24. This view was endorsed and popularized by P. F. Strawson (1952). Pollock defended it in his (1974), but has since rejected it for the reasons described here.

25. Pollock (1984a) has argued that this is indeed the correct explanation for the complexity surrounding inductive reasoning.

examples can be constructed using disjunctions. Suppose, for example, that you would like to confirm inductively that all moose have whiskers. The natural way to proceed would involve collecting a sample of moose and examining them for whiskers. The trouble is, moose are big unruly creatures, and it would be nicer if we could avoid dealing with them. Why not, then, proceed as follows? Consider the disjunctive property of being either a mouse or a moose. We can safely collect a sample of mice-or-moose by just collecting a sample of mice. Upon examining them we find that they are all bewhiskered. That would seem to inductively confirm that everything that is either a mouse or a moose is bewhiskered, and the latter entails that all moose are bewhiskered. So we have a safe way of making inductive generalizations about moose. But obviously, this is absurd. It would be unobjectionable if we could reason inductively about mice-or-moose in the same way we can reason inductively about mice—by collecting an arbitrary sample and generalizing on the basis of it. But we cannot do that. To confirm a generalization about mice-or-moose we must confirm separate generalizations about mice and about moose.[26] To use Goodman's terminology, the property of being either a mouse or a moose is *unprojectible*. Simple rules of inductive reasoning only apply to projectible properties. Thus, in order to give a precise account of inductive reasoning, we need a criterion of projectibility. It has proven remarkably difficult to find such a criterion.

Projectibility is not the only source of difficulty in formulating precise rules of induction. In the case of statistical induction, even the precise form that the conclusion should take is doubtful. If we observe that out of n A's our sample contains m B's, we conclude that the probability of an arbitrary A being a B is *approximately m/n*. That means that the probability lies in some interval around m/n, but how narrow an interval? Untutored intuition does not seem to give us any guidance on this at all.

Another difficulty in formulating rules of induction concerns the circumstances under which inductive reasoning is defeated by peculiarities of the sample. Discovering that the sample is "biased" can disqualify it. It is easy enough to give examples of this phenomenon. For instance, suppose we want to determine the proportion of voters in Indianapolis, who will vote for the Republican gubernatorial candidate in the next election, and we do this by polling a randomly chosen sample of voters. We find that a startling 87 percent of them intend to vote Republican. Prima facie, that gives us a reason for thinking that approximately 87 percent of all voters will vote for the Republican candidate. But if we then discover that, purely by chance, our sample consisted exclusively of voters with incomes greater than $100,000 per year, that would defeat the reasoning. It is easy to illustrate such "fair sample defeaters", but it is much harder to give a general characterization of them. Again, untutored intuition tends to lead us astray.

---

26. For a fuller discussion of projectibility and the unprojectibility of disjunctions, see Pollock (1990).

These difficulties in formulating correct principles of inductive reasoning illustrate that interesting epistemological problems remain even if we dismiss the Humean problem of answering the skeptic. We would like to have an accurate description of the epistemic norms governing induction, and we would like to know whether these norms are fundamental to our system of reasons and reasoning or whether they are derived from simpler and more basic epistemic norms.

5. Theories of Knowledge

The preceding brief discussion of different areas of knowledge illustrates some of the epistemological problems that have excited philosophical interest in those areas. Each area has its own unique problems, and although there may be similarities between the problems that arise in different areas, the differences are as important as the similarities. It is possible to pursue epistemology by focusing on the various areas of knowledge individually and by proceeding piecemeal until each area has been satisfactorily accounted for. This method might yield interesting results, but it is not the primary strategy we will use in this book. This is because there are also more general epistemological problems that arise in all areas of knowledge. These concern the nature and legitimacy of defeasible reasoning, the issue of whether knowledge has "foundations", the source of epistemic norms, and so on. Meeting these challenges will likely provide insight that will be crucial for addressing any of the particular areas of knowledge. We can describe broad categories of epistemological theories in terms of the solutions they propose to these general problems.

Theories of knowledge can be classified in several different ways. First, we can distinguish between "doxastic" and "nondoxastic" theories.

5.1 Doxastic Theories

Until quite recently, it was customarily assumed by epistemologists that the justifiability of a belief is a function exclusively of what beliefs one holds—of one's "doxastic state". To say this is to say that if one holds precisely the same beliefs in two possible circumstances, then no matter how those circumstances differ with respect to things other than what one believes, there will be no difference in what beliefs are justified under those circumstances. We will call this the *doxastic assumption*, and an epistemological theory conforming to this assumption will be called a *doxastic theory*. The doxastic assumption is a very natural one, and no one even considered denying it until fairly recently. The rationale for it is something like the following: all our information about the world is encapsulated in beliefs. It seems that in deciding what to believe, we *cannot* take account of anything except insofar as we have beliefs about it. Consequently, nothing can enter into the determination of epistemic justification except our beliefs. Thus all an epistemological theory can do is tell us how our overall doxastic state determines which of our beliefs can be justified.[27]

The general category of doxastic theories is exhausted by two mutually exclusive subcategories—the foundations theories and the coherence theories:

5.1.1 Foundations Theories

Foundations theories are distinguished by the view that knowledge has "foundations". The contemporary foundations theorist begins with the psychological observation that all knowledge comes to us through our senses. In principle, it is also possible to follow Descartes very closely and attempt to base all knowledge on beliefs that come from pure reason unaided by the senses, but few foundationalists are inclined to defend this view anymore. Foundationalist theories of the 20th century note that our senses provide our only contact with the world around us. Our simplest beliefs about the world are in direct response to sensory input, and then we reason from those simple beliefs to more complicated beliefs (for example, inductive generalizations) that cannot be acquired on the basis of single instances of sense perception. This psychological picture of belief formation suggests a parallel philosophical account of epistemic justification according to which those simple beliefs resulting directly from sense perception form an epistemological foundation and all other beliefs must be justified ultimately by appeal to these *epistemologically basic beliefs*. The basic beliefs themselves are not supposed to stand in need of justification. They are in some sense "self-justifying". One is automatically justified in such a belief merely by virtue of having it. It is typically proposed that epistemologically basic beliefs are beliefs reporting the contents of perceptual states, for example, "There is a red rectangular blob in the upper left hand corner of my visual field."

To complete this picture and build a concrete foundations theory, two things are needed. First, we must have an account of the epistemologically basic beliefs. This must include an account of which beliefs are epistemologically basic, and an account of the sense in which they are self-justifying. Second, we must have an account of "epistemic ascent" —the way in which nonbasic beliefs are justified by appeal to basic beliefs. A number of different answers have been proposed for each of these questions, and they will be examined in detail in the next chapter.

5.1.2 Coherence Theories

What distinguishes foundations theories from other doxastic theories is that they give some limited class of beliefs (the epistemologically basic beliefs) a privileged role in epistemic justification. The basic beliefs justify other beliefs without standing in need of justification themselves. Coherence theories deny that there is any such privileged class of beliefs.

---

27. This objection to nondoxastic theories is raised by Michael Williams (1977). It is also pressed by Laurence Bonjour (1978) (10ff). Ernest Sosa (1981) mentions the objection, but dismisses it.

According to coherence theories, the justifiability of a belief is still a function of one's total doxastic state, but all beliefs are on an epistemological par with one another. This is to characterize coherence theories negatively—in terms of what they deny. Positively, a coherence theory owes us an account of what determines whether a belief is justified. If a belief is not justified by its relationship to a privileged class of basic beliefs, then it must be justified by its relationship to other, ordinary, run-of-the-mill beliefs (after all, that's all there are). Those beliefs are justified by their relationship to further ordinary beliefs, and so on. Instead of justificatory relations being anchored in a foundation, they must meander in and out through our entire network of beliefs. What makes a belief justified is the way it "coheres" with the rest of one's beliefs. Of course, to make this precise we need a precise account of the coherence relation. Different ways of spelling out coherence yield different coherence theories. These theories will be examined in chapter three.

5.2 Nondoxastic Theories

Nondoxastic theories deny the doxastic assumption. Any reasonable epistemological theory will make the justifiability of a belief a function at least partly of what other beliefs one holds, but nondoxastic theories insist that other considerations also enter into the determination of whether a belief can be justified. The naturalness of the doxastic assumption makes it seem initially puzzling how any nondoxastic considerations could be relevant, but one of the main contentions of this book will be that the doxastic assumption is false. It will follow that nondoxastic considerations must be relevant, but it remains puzzling how they can be relevant. Two kinds of answers have been proposed for this question. They are reflected by the internalism/externalism distinction:

5.3 Internalism

Internalism maintains that the justifiability of a belief should be a function of our internal states. Beliefs are internal states, so doxastic theories are internalist theories. Internalists tend to emphasize our conscious internal access to the relations between our beliefs. On this understanding of internalism, reflective, careful agents are able to make epistemological assessments of their beliefs. In *Theory of Knowledge*, Roderick Chisholm writes,

> In making their assumptions, epistemologists presuppose that they are rational beings. This means, in part, that they have certain properties which are such that, if they ask themselves, with respect to any one of these properties, whether or not they have that property, then it will be evident to them that they have it. It means further that they are able to know what they think and believe and that they can recognize inconsistencies. (1989, 5)

Although it is difficult to fill out the details of the kind of conscious access that doxastic internalists favor, the rough idea is that they think of the epistemic agent as possessing a thoroughly penetrating reflective gaze toward her beliefs.

There has been a tendency in the epistemological literature to simply identify internalism with varieties of doxastic internalism, but this identification is too hasty.[28] There are also internalist theories that appeal to more than what we believe, so not all internalist theories are doxastic theories. To see this, it will be useful for us to make some preliminary remarks about our own epistemological views. Foundationalism takes as its starting point the observation that our knowledge of the world comes to us through perception, broadly construed, and attempts to accommodate that by positing the existence of self-justifying epistemologically basic beliefs reporting our perceptual states. We will argue that all foundations theories are false for the simple reason that people rarely have any epistemologically basic beliefs, and never have enough to provide a foundation for the rest of our knowledge. That can be taken to motivate coherence theories, which give no special place to beliefs pertaining to perception. But we will argue that all coherence theories fail for a related reason—they are unable to accommodate perception as the basic source of our knowledge of the world. We claim that in determining whether a belief is justified, importance must be attached to perceptual states, but this cannot be accomplished by looking at *beliefs about* perceptual states. This suggests that justification must be partly a function of the perceptual states themselves and not just a function of our beliefs about the perceptual states. This sort of view is called *direct realism*, and a version of direct realism will be defended in this book. That is why, in our taxonomy, there can be nondoxastic internalist theories.

The idea behind internalism is that the justifiedness of a belief is determined by whether it was arrived at or is currently sustained by "correct cognitive processes". The view is that being justified in holding a belief consists of conforming to epistemic norms, where the latter tell you "how to" acquire new beliefs and reject old ones. In other words, epistemic norms describe which cognitive processes are correct and which are incorrect, and being justified consists of "making the right moves". Internalist theories are committed to the principle that the correctness of an epistemic move (a cognitive process) is an inherent feature of it. For example, it may be claimed that reasoning in accordance with *modus ponens* is always correct, whereas arriving at beliefs through wishful thinking is always incorrect. This is implied by the claim that the justifiability of a belief is a function of one's internal states, because what that means is that we can vary everything about the situation other than the internal states without affecting which beliefs are justifiable. In particular, varying contingent properties of the cognitive processes themselves will not affect whether a belief is justified. This is called *cognitive essentialism*. According to cognitive essentialism, the epistemic correctness of a cognitive process is an essential feature of that process and is not affected by contingent facts such as the reliability of the process in the actual world.

---

28. For sensitive discussions of the varieties of internalism, see William Alston (1986), Richard Fumerton (1995), and Alvin Plantinga (1993a).

5.4 Externalism

Externalism is the denial of internalism. According to externalism, more than just the internal states of the believer enter into the justification of beliefs. A wide variety of externalist theories are possible. What we might call *process externalism* agrees with the internalist that the epistemic worth of a belief should be determined by the cognitive processes from which it issues, but it denies cognitive essentialism according to which the correctness of a cognitive process is an essential property of it. It insists instead that the same cognitive process could be correct in some circumstances and incorrect in others. A view of this sort is represented by the *process reliabilist* who proposes that cognitive processes should be evaluated in terms of their reliability in producing true beliefs.[29]

Reliabilist theories stand in marked contrast to more traditional epistemological theories. The reliability of a cognitive process is a contingent matter. For example, a cognitive process on which we place great reliance is color vision. Color vision is reasonably reliable in the normal environment of earth-bound human beings. But if we lived in an environment in which the colors of our light sources varied erratically, color vision would be unreliable. So, the correctness of the cognitive processes involved in color vision is not intrinsic to color vision per se, but rather is a function of the relationship between the process and the environment. Giving a precise specification of which environments are relevant in assessing a cognitive process is a difficult matter. This has led some epistemologists to modify reliabilism by proposing a theory of proper functions.[30] The theory of proper functions holds that a belief is justified in case it is the product of a process that is working according to its proper function in the environment for which it is appropriate. The theory of proper functions is similar to process reliabilism in maintaining that the proper function of a cognitive mechanism will be aimed at reliability. Whereas process reliabilism does not make reference to the design plan behind the cognitive process, the theory of proper functions stipulates that the process has to offer its output in accordance with a design plan. Neither the reliability of a cognitive process nor whether a cognitive process is functioning properly can be assessed a priori. These depend upon contingent matters of fact. Thus reliabilism and the theory of proper functions make epistemic justification turn on contingent matters of fact. Cognitive essentialism is false on this view.

A different kind of externalist theory is *probabilism*, which assesses beliefs in terms of their probability of being true. Probabilism makes no explicit appeal to the cognitive pedigree of a belief, although the probability of a belief being true can of course be indirectly influenced by the cognitive processes from which it derives. Probabilism has been quite influential in the philosophy of science, where it is part of what is called "Bayesian

---

29. The best known of these is due to Alvin Goldman (1979) and (1986).
30. See Alvin Plantinga (1988) and (1993b).

epistemology". It has had little influence on epistemology outside of the philosophy of science, but it deserves a careful discussion and will be treated at length in chapter four.

One of the attractions of externalist theories is that they hold out promise for integrating epistemic norms into a naturalistic picture of human beings. As we noted at the beginning of this chapter, contemporary philosophers have been attracted by the conception of human beings as creatures in the world—biological machines that think. Epistemic norms should emerge from this psychological construction, but their very normativity has seemed to make them resistant to such an account. Philosophers raised on the naturalistic fallacy in ethics are prone to suppose that naturalistic theories of normative concepts are impossible. But if epistemic relations can be reduced to considerations of probability or reliability, or to some other naturalistic concept, this obstacle dissolves. Externalist theories have seemed to provide the only possible candidates for naturalistic reductions of epistemic norms, so this has made them attractive in the eyes of many philosophers.

Externalist theories are automatically nondoxastic theories. That is, they take the justifiability of a belief to be a function of more than just one's total doxastic state. This will prove to be a source of difficulty for externalist theories. We will argue that nondoxastic internalist theories can escape the objection that in deciding what to believe, you cannot take account of anything except insofar as you have beliefs about it. They can do that by maintaining that in the requisite sense you can take account of other internal states. But you cannot similarly take account of external states, and that will prove to be the ultimate downfall of externalism. This cannot be argued convincingly until chapter five.

5.5 Plan of the Book

The categories of epistemological theories are related to each other as follows:

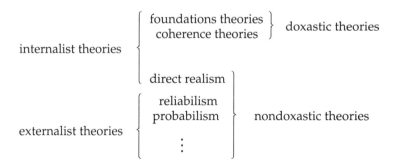

There are still many subtleties to address and issues that need to be made more precise. The plan of the rest of this book is as follows: Chapter two will attempt to construct the most plausible kind of foundations theory possible. Even though some crucial groundwork for our own view of reasoning will emerge here, the ultimate conclusion of chapter two will be that all foundations theories are false. If all foundations theories are false but the doxastic assumption is true then some coherence theory must be true. Chapter three will take up the general discussion of coherence theories, distinguishing between a number of different varieties, narrowing the class of feasible candidates to just a few and then rejecting them all. It follows that all doxastic theories are false.

The way in which those theories fail will suggest that the solution may lie in direct realism, and that theory will be sketched at the end of chapter three. Its defense, however, will await chapters five and six. Chapter four will discuss several kinds of externalist theories in detail. Reasons will be given for regarding externalist theories as an attractive alternative to doxastic theories, but detailed objections will also be raised to existing externalist theories. These objections will dispose of most externalist theories that have actually been proposed in the literature, but the possibility will remain that some other kind of externalist theory might be true. Chapter five will attempt to tie it all together. Chapter five will explore the nature of epistemic norms, and will use the resulting account to propound a general refutation of doxastic theories and externalist theories.

The intuition behind the doxastic assumption is that in deciding what to believe, we cannot take account of anything except insofar as we have beliefs about it. That has to be wrong because the only kinds of doxastic theories are foundations theories and coherence theories, and they are all false. Explaining how the doxastic assumption goes wrong will occupy a large part of chapter five. We will not try to give the details here, but the general idea will be that we do not literally "decide" what to believe. That is to over-intellectualize the process of belief acquisition. Belief acquisition is determined by cognitive processes that have access to more than just our beliefs. Beliefs and perceptual states are alike in being "internal states". These are, roughly, states of ourselves to which we have "direct access", and our cognitive processes can appeal to internal states in general—not just to beliefs.

It will be urged in chapter six that a proper understanding of epistemic norms makes them amenable to a naturalistic account that is not an externalist account. We will thus be led to a kind of naturalistic internalism. Chapters five and six form the core of our understanding of human beings as cognizers capable of acquiring all the diverse kinds of knowledge that can, ultimately, lead to the discovery of DNA or a cure for cancer.

The conclusions of chapters five and six are still very general. They support a high-level conclusion to the effect that *some* form of direct realism is true, but leave the details undetermined. For this high-level account of rational cognition to be correct, it must be possible to construct a version of direct realism that can accommodate all the different kinds of knowledge discussed in section four. Chapter seven will give a brief presentation of a direct realist theory that aims to accomplish this.

2
FOUNDATIONS THEORIES

1. Motivation

Until quite recently, the most popular epistemological theories were all foundations theories.[31] Foundations theories are distinguished from other doxastic theories by the fact that they take a limited class of "epistemologically basic" beliefs to have a privileged epistemic status. It is supposed that basic beliefs do not stand in need of justification—they are "self-justifying". Nonbasic beliefs, on the other hand, are all supposed to be justified ultimately by appeal to basic beliefs. Thus the basic beliefs provide a foundation for epistemic justification.

The simple motivation for foundations theories is the psychological observation that we have various ways of sensing the world, and all knowledge comes to us via those senses. The foundationalist takes this to mean that our senses provide us with what are then identified as epistemologically basic beliefs. We arrive at other beliefs by reasoning (construed broadly). Reasoning, it seems, can only justify us in holding a belief if we are already justified in holding the beliefs from which we reason, so reasoning cannot provide an ultimate source of justification. Only perception can do that. We thus acquire the picture of our beliefs forming a kind of pyramid, with the basic beliefs provided by perception forming the foundation, and all other justified beliefs being supported by reasoning that traces back ultimately to the basic beliefs.

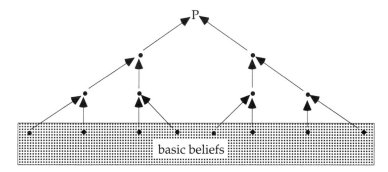

basic beliefs

---

31. Examples of foundations theories can be found in Rudolf Carnap (1967), C. I. Lewis (1946), Nelson Goodman (1951), Roderick Chisholm (1966), (1977), and (1981), Pollock (1974), and Paul Moser (1985). Foundations theories may be experiencing a renaissance. See Michael DePaul (1999).

The foundationalist picture seems to derive rather directly from psychological truisms, and this gives it considerable force. But the picture must be filled out in two respects before we have anything that deserves to be regarded as a concrete epistemological theory. First, we must know more about the basic beliefs. What kinds of beliefs are basic, and what suits them for such a privileged role? Second, we must know more about the way nonbasic beliefs are supported through reasoning from the basic beliefs. These two broad topics will be the subjects of the next two sections. The problem of reasoning will prove more subtle than foundationalists often suppose, so in section four we attempt to develop a more nuanced foundationalism.

2. Basic Beliefs

It is the existence of epistemologically basic beliefs that distinguishes foundations theories from coherence theories (recall from chapter one that coherence theories do not accord a special epistemic status to any privileged subset of the entire doxastic corpus). Basic beliefs must be justified independently of reasoning—if a belief can only be justified through reasoning, its justification is dependent on the justification of the beliefs from which the reasoning proceeds, and hence, by definition, it is not a basic belief. It seems that the only beliefs that are not held on the basis of reasoning are those held directly on the basis of various kinds of perception. Thus basic beliefs must be perceptual beliefs, in a sense yet to be made precise. But if basic beliefs are to provide a foundation, they must themselves have a secure epistemic status. Can such a status be granted to perceptual beliefs?

The beliefs we normally regard as the immediate result of perception are beliefs about physical objects. For example, I see that the door is open, I hear someone climbing the stairs, I smell the fish frying, I feel (proprioceptively sense) that my fist is clenched, and so on. An undeniable feature of such beliefs is that they can be mistaken. At the very least, we can be fooled by unusual perceptual environments or by perceptual illusions. For example, if I see a red shirt in green light I may think it is black. Furthermore, our perceptual beliefs are strongly influenced by our expectations. If I expect to see my brother sitting behind his desk, I may (at least momentarily) think I do see him if his chair is occupied by someone who resembles him only vaguely. There are many possible sources of perceptual error. Ordinary perceptual beliefs cannot be taken for granted. If they are fallible, it seems that they stand as much in need of justification as any other beliefs, and hence cannot provide the stopping point for justificatory appeals. How, then, can perceptual beliefs be basic?

The traditional response of the foundationalist has been to deny that basic beliefs are *ordinary* perceptual beliefs. Instead, the foundationalist retreats to a weaker kind of belief. Basic beliefs must be perceptual beliefs *in some sense*, but they need not be beliefs about physical objects. I can be mistaken about what color something is, but it is not so obvious that I can be mistaken about what color it *looks to me*. More generally, in perception I have sensory experiences, and these sensory experiences lead me to have beliefs about my physical surroundings. I can be wrong about those surroundings, but can I be wrong about the character of my sensory experiences? If not, we might reasonably regard beliefs about the latter as basic, and take beliefs about physical objects to be supported only indirectly, by reasoning from beliefs about our sensory experiences.

It is useful to have some convenient terminology for describing the character of our sensory experiences. An artificial terminology that has acquired some currency in contemporary epistemology is the "appeared to" terminology. If, associated with a state of affairs *P*, there is a kind of sensory experience one standardly has when one is in that state of affairs, we can refer to that sensory experience as *being appeared to as if P*. Thus I can talk about being appeared to as if there is something red before me. Sometimes there will be a convenient adverbial locution that will allow us to shorten the description and say things like, "I am appeared to redly". A few philosophers have advanced complicated theoretical rationales for adopting this way of describing the character of sensory experience,[32] but we do not mean to endorse any such rationale here. For us, it is just a convenient way of talking. The point of such strange adverbial constructions is to make clear the distinction between ordinary perceptual beliefs about the way the world is and the special kind of belief that only ventures a claim about how things appear.

The suggestion is that basic beliefs are beliefs about ways of being appeared to—"appearance beliefs" for short. This is motivated by the observation that the only other candidates for epistemologically basic beliefs—perceptual beliefs about physical objects—can be mistaken and thus seem to stand in need of justification themselves and cannot be epistemologically basic. This motivation needs to be examined more carefully. It involves the presumption that if a belief can be mistaken then it is not a basic belief, and also the presumption that appearance beliefs cannot be mistaken. Either of these presumptions could be denied.

For a foundations theory to work, the class of basic beliefs must satisfy two conditions: (1) there must be enough basic beliefs to provide a foundation for all other justified beliefs, and (2) the basic beliefs must have a secure status that does not require them to be justified by appeal to further justified beliefs. We have formulated this second condition by saying that basic beliefs must be *self-justifying* in the sense that one can be justified in holding such a belief merely by virtue of the fact that one does hold it—one does not need an independent reason for holding a basic belief. The concept of a self-justifying belief can be made more

---

32. See Roderick Chisholm (1957) and Michael Tye (1984).

precise in either of two ways. First, consider the simplest concept of self-justification, that of an "incorrigibly justified" belief:

DEFINITION:
A belief is *incorrigibly justified* for a person S if and only if it is impossible for S to hold the belief but be unjustified in doing so.

Few beliefs are incorrigibly justified. It is quite possible to believe, for no reason at all, and in the face of considerable evidence to the contrary, that there is a book before me. It would be unusual to hold that belief under those circumstances, but it would not be impossible, and if I were perverse enough to do so then the belief would not be justified. Thus the belief that there is a book before me is not incorrigibly justified. Similar reasoning will establish that most beliefs fail to be incorrigibly justified, but perhaps a few will slip through the net of the argument. Among these might be appearance beliefs. We will return to this question shortly.

If basic beliefs are incorrigibly justified then they provide the firmest possible foundation for the justification of other beliefs. That is the attraction of the concept of incorrigible justification. But incorrigible justification is more than is needed for basic beliefs. Basic beliefs are beliefs from which justification starts, and thus it must be *possible* to be justified in holding a basic belief without having a reason for it, but it need not be the case that one is *always* justified in holding such a belief. There must be a presumption in favor of justification, so that in the face of no counter-evidence one is justified in holding a basic belief, but this does not preclude the possibility that one's justification can be defeated by appropriate counter-evidence. In other words, one must be able to hold basic beliefs justifiably without having reasons for them, but reasons could still be relevant in a negative way by making one unjustified in holding such a belief when she has a reason for thinking it false. This is captured by the following definition:

DEFINITION:
A belief is *prima facie justified* for a person S if and only if it is only possible for S to hold the belief unjustifiedly if she has reason for thinking she should not hold the belief (equivalently, it is necessarily true that if S holds the belief and has no reason for thinking she should not then she is justified in holding the belief).[33]

When we have a belief, there is something that we believe. Objects of

---

33. This definition must be supplemented with an explanation of what counts as a reason for thinking S should not hold a belief P. At the very least, anything that is a reason for S to believe ~P despite the fact that she currently believes P is such a reason. More generally, we would suggest that something is a reason for S to think she should not believe P just in case, when conjoined with the fact that S does believe P it yields a reason for doubting that S would not believe P unless P were true.

belief are called "propositions". For example, when I believe that there is a cat in the window, the object of my belief is the proposition that there is a cat in the window.[34] The nature of propositions is a matter of great philosophical dispute, and for that reason we tried to avoid talking about them in writing this book. But that proved stylistically awkward, so we have capitulated and will make reference to propositions in talking about beliefs. As we use the term, propositions are just possible objects of belief. We make no substantive assumptions about the nature of propositions here, and we believe that at considerable stylistic expense, all reference to propositions in this book could be replaced by talk of "possible beliefs". Having said that, we will now allow ourselves free use of propositions. We will call the object of an epistemologically basic belief "an epistemologically basic proposition", and we will say that a proposition is incorrigibly justified or prima facie justified just in case belief in that proposition would be incorrigibly justified or prima facie justified.

We have made the abstract point that if basic beliefs are to provide a foundation for knowledge then they must be self-justifying, in the sense either of being incorrigibly justified or prima facie justified. But is there any reason to expect some beliefs to have such a status? What could make a belief self-justifying? The most common answer has been that some beliefs *cannot be mistaken*—if you hold such a belief then it follows logically that the belief is true. Such beliefs are said to be "incorrigible". The most common definition of incorrigibility is as follows:

PROVISIONAL DEFINITION:
A belief is *incorrigible* for a person S if and only if it is impossible for S to hold the belief and be wrong.

Ordinary beliefs are not incorrigible. Just as I can be unjustified in believing that there is a book before me, it is quite possible for me to believe that there is a book before me when there is none, so my belief that there is a book before me is neither incorrigibly justified nor incorrigible. It has seemed to many philosophers, however, that appearance beliefs are incorrigible. It has been urged that such beliefs cannot be mistaken. The simple motivation for this claim lies in the difficulty of imagining what could possibly show that one is wrong in thinking, for instance, that he is appeared to redly. It seems that other people can have various kinds of evidence for thinking that I am or am not appeared to redly, but that evidence cannot be relevant for me. *I* can tell, just by reflecting on the matter, whether I am appeared to redly.

---

34. Part of the reason philosophers wish to detach the object of a belief from the belief itself is that it seems we can have a different mental relation than belief to the same object. Contrast the belief that there is a cat in the window with the desire or hope that there is a cat in the window. The object of the desire or hope might be the same as the object of the belief, even though desire and hope differ from belief.

People are persuaded by this kind of reasoning, but it does not constitute an argument. It really amounts to no more than an assertion that
such beliefs are incorrigible. Nevertheless, it illustrates the intuitive
appeal that that idea has. Regardless of whether we ultimately decide
that appearance beliefs are incorrigible, one must acknowledge and be
prepared to explain the intuitive pull of that idea.

The methodological attractiveness of incorrigibility is that it seems to
offer an explanation for how basic beliefs could be self-justifying. If a
belief cannot possibly be false, then it is apt to seem that you have the
best possible justification for holding it. What more could you want?
But this is misleading. It is now generally recognized that the above
definition of incorrigibility includes beliefs that we do not want to regard
as self-justifying. The definition is too permissive. Consider any necessary
truth P. For instance, P might be some complicated theorem of mathematics. Now consider a student who is trying to solve a problem on an
exam. She has never seen this theorem, has no reason to believe it, and
in fact the theorem is counter-intuitive. But it occurs to her that if the
theorem were true then she could solve her problem, so wishful thinking
leads her to believe it. Clearly, she is not justified in this belief. A
person is not automatically justified in believing a mathematical truth if
she believes it for no reason. We need reasons for believing principles of
mathematics just as much as we do for believing contingent truths. But
it is impossible to believe a mathematical truth and be wrong. Because
mathematical truths are necessarily true, it is also necessarily true that if
one believes a mathematical truth then one is right. So by the provisional
definition, belief in any mathematical truth is incorrigible.

Incorrigibility, defined as above, does not guarantee that a belief is
self-justifying. The difficulty is that the definition does not capture the
idea of it being impossible for us to be mistaken about P. We want it to
be impossible for us to be mistaken about *whether* P is true. In other
words, believing P should guarantee that P is true, and believing $\sim P$
should guarantee that P is false. Thus we propose to redefine incorrigibility
as follows:

DEFINITION:
A proposition P is *incorrigible* for a person S if and only if (1) it is
necessarily true that if S believes P then P is true, and (2) it is
necessarily true that if S believes $\sim P$ then P is false.

This defines incorrigibility for propositions. We will say that a belief is
incorrigible just in case it is belief in an incorrigible proposition. This
definition captures the idea that belief about whether P is true is a conclusive arbiter of whether P is true. Mathematical truths are not incorrigible
in this sense because they remain true even if one disbelieves them. It
remains at least somewhat plausible that appearance beliefs are incorrigible, and it seems reasonable that incorrigibility in this strong sense is
sufficient to guarantee incorrigible justification.

Foundationalism requires that there be self-justifying epistemologically basic beliefs. We have not resolved the question whether there are
any self-justifying beliefs, but we have clarified the logical geography of
the concept of self-justification and prepared the way for a more definitive
discussion that will occur in section four. Before undertaking that discussion, however, it is convenient to investigate the matter of epistemic
ascent.

3. Epistemic Ascent

Even if the foundationalist could secure a fund of epistemologically
basic beliefs, the problem would remain of getting from them to other
justified beliefs. Nonbasic beliefs are justified by reasoning from basic
beliefs. Foundational theories owe a detailed account of reasoning. In
this section we will fill out some of that account.

Reasoning proceeds in terms of reasons. We can define:

DEFINITION:
A belief P is a *reason* for a person S to believe Q if and only if it is
logically possible for S to become justified in believing Q by believing it on the basis of P.

This definition appeals to the psychological relation of holding one belief
on the basis of another. This is called *the basing relation*. The basing
relation is important in epistemology. To be justified in believing something it is not sufficient merely to *have* a good reason for believing it.
One could have a good reason at one's disposal but never make the
connection. Suppose, for instance, that you are giving a mathematical
proof. At a certain point you get stuck. You want to derive a particular
intermediate conclusion, but you cannot see how to do it. In despair,
you just write it down and think to yourself, "That's got to be true." In
fact, the conclusion follows from two earlier lines by *modus ponens*, but
you have overlooked that. Surely, you are not justified in believing the
conclusion, despite the fact that you have impeccable reasons for it at
your disposal. What is lacking is that you do not believe the conclusion
on the basis of those reasons.

Although the basing relation is of manifest importance to epistemology,
it is difficult to say much about it in an a priori way. It is in some loose
sense a causal relation, but the mere fact that holding one belief causes a
person to hold another is not sufficient to guarantee that he holds the
second belief on the basis of the first.[35] Our beliefs can be tied together

---

35. For an alternative view, see Keith Lehrer (1971) and (1990), and Richard
Feldman and Earl Connee (1985), who argue that the basing relation is not causal.
When we deal with coherentism in the next chapter, we will offer grounds for rejecting

by all sorts of aberrant causal chains. I might believe that I am going to be late to my class, and that might cause me to run on a slippery sidewalk, lose my footing, and fall down, whereupon I find myself flat on my back looking up at the birds in the tree above me. My belief that I was going to be late to class caused me to have the belief that there were birds in that tree, but I do not believe the latter on the basis of the former. Giving an informative philosophical analysis of the basing relation is what has come to be called *the problem of the basing relation*. At this point it is hard to say anything helpful about it, but when we return to this problem in chapter five we will be able to make some progress with it.[36]

So, in the foundationalist's picture, nonbasic beliefs are justified on the basis of reasons. What kinds of reasons are there?

3.1 Defeasible Reasons

In chapter one, when we discussed Hume's skeptical argument regarding induction, we encountered the assumption that a reason can only be a good reason for believing its conclusion if it logically entails that conclusion. That has been a common assumption in the history of epistemology. A frequently encountered variant of it has been that reasons must be either entailments or inductive reasons. We feel that one of the most important advances of contemporary epistemology has been the rejection of both of these assumptions and the recognition of reasons that are neither inductive reasons nor logical entailments.[37]

We shall call a reason that logically entails its conclusion a *conclusive* reason. Inductive reasons are nonconclusive reasons, and it will be argued below that there are many other kinds of nonconclusive reasons as well. Let us begin by exploring the characteristics of nonconclusive reasons, taking induction to be a paradigm of such a reason. The most important characteristic of nonconclusive reasons is that they are *defeasible*. For instance, inductive evidence creates a rational presumption in favor of a generalization, and in the absence of any other relevant information it can justify belief in the generalization, but the presumption can be *defeated*

---

non-causal accounts of the basing relation.

36. See George Pappas (1979) and William Alston (1988) for other discussions of the basing relation.

37. The two contemporary epistemologists who have made the most of this are Roderick Chisholm ((1966), (1977), and (1981)), and John Pollock (originally in his PhD dissertation at Berkeley in 1965, then in (1967), (1970), and (1974)). Pollock was primarily influenced by the rather sketchy remarks about "criteria" in Wittgenstein (1953). It is interesting that defeasible reasoning was discovered independently a few years later by researchers working on artificial intelligence, and has been the subject of considerable research in that field. They use the terms "default reasoning" and "non-monotonic reasoning" for defeasible reasoning. The original papers were those of Jon Doyle (1979), Raymond Reiter (1978) and (1980), Drew McDermott and Jon Doyle (1980), and John McCarthy (1980). A general overview of early AI work on non-monotonic reasoning can be found in Matt Ginsberg (1987). A survey of recent work on defeasible reasoning in AI can be found in H. Prakken and G. Vreeswijk (2000).

by various kinds of considerations. Most simply, if we know of lots of *A*'s and they are all *B*'s, that can justify us in believing that all *A*'s are *B*'s, but if we subsequently encounter even a single *A* that is not *B*'s, all of the previous evidence counts for nothing toward the proposition that all *A*'s are *B*'s. We are still justified in believing the evidence that constituted our original reason, but now we have further information that constitutes a *defeater*. Precisely:

DEFINITION:
If *P* is a reason for *S* to believe *Q*, R is a *defeater* for this reason if and only if (*P*&R) is not a reason for *S* to believe *Q*.

Defeasible reasons are reasons for which there can be defeaters. Such a reason is called a *defeasible reason*.

3.2 Justified Belief and Undefeated Arguments
The vehicle for epistemic ascent is reasoning. Reasoning proceeds by stringing reasons together into arguments. We can think of an argument as a finite sequence of propositions ordered in such a way that for each proposition *P* in the sequence, either (1) *P* is epistemologically basic or (2) there is a proposition (or set of propositions) earlier in the sequence that is a reason for *P*.[38] A person *instantiates* an argument if and only if she believes the propositions comprising the argument and she believes each nonbasic proposition in it on the basis of reasons for it that occur earlier in the argument. Let us say that an argument *supports* a proposition if and only if that proposition is the final proposition in the argument.

What is the connection between arguments and justified belief? It might be supposed that a foundationalist should take belief in *P* to be justified for a person *S* if and only if *S* instantiates an argument supporting *P*.[39] But this simple proposal fails because it overlooks defeasibility. To illustrate, suppose a person *S* simultaneously instantiates arguments of each of the forms shown in figure 2.1. The arrows in the figure represent the basing relation in the arguments. The conclusion of the second argument is a defeater for the first argument. Under the circumstances, it seems that *S* is not justified in believing that *x* is red. Of course, that could change if *S* also instantiates an argument supporting a defeater for the second argument. That would reinstate the first argument.

---

38. This is probably simplistic in at least one respect. It seems likely that we should allow arguments to contain subsidiary arguments. For example, an argument might contain a subsidiary argument supporting a conditional by conditional proof, or a subsidiary argument supporting a negation by *reductio ad absurdum*. The additional sophistications that this requires are not relevant to the present discussion, so we will ignore them for now.

39. To accommodate belief in epistemologically basic propositions, we can think of such a proposition in isolation as comprising a one-line argument and take this as the limiting case of an argument supporting a proposition.

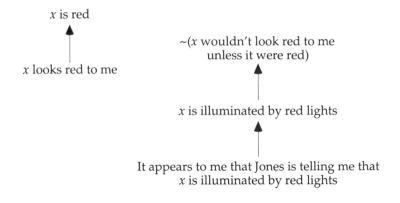

Figure 2.1 Defeat of one argument by another

To handle defeasibility in a general way, we must recognize that arguments can defeat one another and that a defeated argument can be reinstated if the arguments defeating it are defeated in turn. We can then understand the foundationalist as asserting that belief in P is justified for a person S if and only if S instantiates an undefeated argument supporting P.

3.3 The Problem of Perception

We have before us a plausible reconstruction of epistemic ascent in foundationalism. Still, the discussion of ascent has proceeded on a fairly abstract level, and the only examples of nonconclusive reasons we have given are inductive reasons. A defensible foundationalism will maintain that there are other nonconclusive reasons as well. Consequently, we shall take a short detour through a detailed appraisal of epistemic ascent in another area that is of some importance to epistemologists. We will do this by focusing on the problem of perception. This is the problem of explaining how we can acquire knowledge of our physical surroundings through sense perception. For instance, we might judge that an object is red because it looks red to us. What justifies our reasoning in this way?

3.3.1 Phenomenalism

Although we have rejected the idea that the only reasons are conclusive reasons, that supposition played an important historical role in the problem of perception. On that supposition, we could only explain perceptual knowledge by finding logical entailments between the way things look and the way they are. To get from an object's looking red to its being red we would have to find some further premise which, when conjoined with the fact that the object looks red to us, entails that the object is red. Epistemologists in the first half of the twentieth century supposed that the only way such an entailment could arise is from a logical analysis of what it is for an object to be red, where that analysis proceeds in terms of appearances.[40] They believed this because they adopted a particular view of concepts according to which concepts are characterized by definitions stating necessary and sufficient conditions for something to exemplify them. Such definitions state the *truth conditions* of the concepts. It was supposed that, with the exception of some "logically simple" concepts that do not have definitions, the logical nature of a concept is completely given by its definition and all logical properties of a concept must emerge from its definition. In particular, entailments between concepts must arise from the definitions of those concepts. Thus, if there is to be an entailment between something's looking red under some circumstances *C* and its being red, this must result from the definition of either the concept of something being red or the concept of its looking red. The latter was generally accepted as one of the logically simple concepts in terms of which definitions must ultimately be framed, so it was assumed that the requisite entailments must arise from a correct analysis of the concept of something being red.

The proposal was to solve the problem of perception by finding entailments of the form:

x's looking red to *S* under circumstances of type *C* entails that *x* is red.

If circumstances of type *C* involve reference to the states of material objects, then this will not solve the problem of perception because it will presuppose some perceptual knowledge in order to obtain other perceptual knowledge. Consequently, circumstances of type *C* can only involve reference to appearances. As such an entailment was supposed to arise out of the definition of 'red', it followed that there had to be a definition of 'red' that proceeded entirely in terms of appearances. The view that the problem of perception can be solved by finding such definitions is called *phenomenalism*. Phenomenalism was once the predominant theory in epistemology, not because philosophers had concrete phenomenalist analyses to actually propose, but because reasoning like the above convinced them that phenomenalism was the only possible solution to the problem of perception.

Phenomenalism is apt to seem antecedently preposterous. What kind of analysis of material object concepts could possibly be proposed in terms of ways of being appeared to? But phenomenalists had some ingenious proposals to make in this connection. The most sophisticated phenomenalist analysis was that proposed by C. I. Lewis (1946). His suggestion was that "*x* is red" can be analyzed as a conjunction of possibly

---

40. See, for example: Rudolf Carnap (1967), C. I. Lewis (1946), and Nelson Goodman (1951).

infinitely many conditionals of the form "if I were to do *A* in circumstances *C* then I would be appeared to R'ly", where R describes some way of being appeared to. For example, such a conditional might tell us that if we look at *x* under normal lighting conditions then it will look red to us.

There are a number of problems with such a phenomenalist analysis. The most serious difficulty concerns the circumstances *C* to which the conditionals appeal. The example was "lighting conditions are normal", but that will not do because these conditionals must be formulated entirely in terms of appearances without making appeal to the states of material objects. The circumstances must be "phenomenal" circumstances in the sense that they are formulated entirely in terms of appearances. If we look seriously for phenomenal circumstances that might do the job, no plausible candidates come to mind. Furthermore, there is a simple argument that seems to show that there can be no phenomenal circumstances that can play the requisite role in a phenomenalist analysis. This is due to Roderick Firth (1950), and it is called *the argument from perceptual relativity*. The argument proceeds by noting that being in circumstances of type *C* must logically entail that if *x* is red and I look at it then it will appear red to me. But there can be no phenomenal circumstances that have this entailment. No matter what we propose for *C*, we can always imagine elaborate conditions under which the putative entailment will be falsified. For example, a person might be wired into a computer that interferes with her perceptual inputs selectively, so that it does not prevent normal perception in most cases, but it does regulate her perceptual experience to the extent of making phenomenal circumstances *C* hold no matter what is going on around her. If we then place her in a standard sort of situation in which red things fail to look red (e.g., illumination by green light), she may see a red object and it may fail to look red to her. Thus it seems that no such phenomenalist entailments can hold, and hence a phenomenalist analysis is impossible.

3.3.2 Using Induction

The main attraction of phenomenalism was that it offered the prospect of explaining perceptual knowledge within a framework that recognized only conclusive reasons. But once it is acknowledged that at least inductive reasons are nonconclusive reasons, there is little reason for wanting to confine ourselves to such a framework. This suggests that perceptual knowledge might be accommodated by various kinds of inductive reasoning. The simplest kind of inductive reasoning would involve discovering that objects that look red usually are, or perhaps that objects that look red under specifiable circumstances are always red. Then noting that something looks red (under the appropriate circumstances) would give us a reason for thinking it is red. But this kind of reasoning cannot legitimate perceptual knowledge because we could only make such discoveries if we had independent access to the colors of objects and the colors they look to us and could then compare them. We have no ultimate access to the colors of things except via how they appear to us, and it is the legitimation of that inference that is in question in the problem of perception. Thus we would have to already solve the problem of perception before we could confirm such inductive generalizations.

3.3.3 Inference to the Best Explanation

There is, however, another form of inductive reasoning that has seemed to hold out more hope. A kind of induction that is common in everyday life and, in some form, underlies much of science, is what has come to be called "inference to the best explanation".[41] Given a set of observations, we often take a hypothesis to be confirmed because it is the best explanation of those observations. For example, if I see dust in the air and the limbs of the trees swaying about outside the window I may infer that it is windy because that is the best explanation for what I see. Similarly, a physicist may infer that most elementary particles are composed of quarks on the grounds that that best explains the interrelationships that have been observed between elementary particles. The confirmation of scientific theories is probably best viewed in terms of inference to the best explanation. There are lots of problems concerning how this form of induction is to be analyzed and made precise, but let us waive those problems for now. It cannot reasonably be denied that they have solutions even if we are not yet in a position to give them.

It looks at first as though it is very easy to reconstruct perceptual knowledge in terms of inference to the best explanation. Under ordinary circumstances, surely the best explanation for why something looks red to us is that it is red! But before we endorse this account too hastily, note that there is a distinction between comparative and noncomparative appearance judgments.[42] Comparative appearance judgments classify the way one is appeared to as being the way one is normally appeared to under some objectively describable circumstances. It seems that the very possibility of comparative appearance judgments presupposes our ability to make noncomparative appearance judgments. In order to know that I am appeared to in the way I am normally appeared to when I see something red, I must be able to tell how I am appeared to, I must know (inductively) how I am normally appeared to when I see something red, and then I must judge that these are the same. Being able to tell how I am appeared to, in this sense, is to be able to make a noncomparative judgment. If "looks red" is being used comparatively, then it seems right that the best explanation for an object's looking red is ordinarily that it is red. But comparative judgments about how things look are not epistemologically basic because they must be based on prior beliefs about, for example,

---

41. See particularly Gilbert Harman (1973) and (1980), and Peter Lipton (1991).
42. Roderick Chisholm (1957).

how red things normally look. Accordingly, this does not secure epistemic ascent from basic beliefs. We must instead focus on noncomparative appearance judgments. If "looks red" is being used noncomparatively, then there is no obvious connection between the concept of looking red and the concept of being red. "That looks red to me" seems to amount to thinking about a particular way things can look and then thinking, "This looks *that way* to me". But why should something's being red count as an explanation for its looking that way to me? Its being green would be just as much of an explanation, which is to say that neither is any explanation at all. More accurately, the object's being red only explains its looking the way it does to me if I already know that that is the way red things ordinarily look to me, but of course, I cannot know that without first acquiring some perceptual knowledge. So despite initial appearances, perceptual knowledge cannot be modeled on this sort of use of inference to the best explanation.

3.3.4 Scientific Realism

There is another way to approach the problem of perception by using inference to the best explanation. According to what is sometimes called scientific realism, we posit the existence of the physical world as the best explanation for why things appear to us as they do.[43] This differs from the previous use of inference to the best explanation by taking perception to involve inference to a global theory of the world rather than just local inferences to, for instance, the color of a particular object. By thus changing the scope of the theory it avoids the earlier logical difficulties. For example, rather than inferring that something is red because it looks red, this account would have us infer much more generally that there is a way things tend to be when they look red to us, and we call that way "being red". But the cost of avoiding the logical difficulties is to make the account subject to overwhelming psychological difficulties. The perception of objects is a largely automatic process that does not involve anything like deliberate postulation for the sake of explanation. The main place this makes a difference is to the amount of evidence that is required for a reasonable perceptual judgment. According to scientific realism we are at each point confirming a global theory of the physical world, so before we can even get started reasoning this way we must acquire a great deal of information about how things appear to us on various occasions and what regularities there are in how things appear to us. In fact, however, we rarely take note of how things appear to us, and we almost never remember enough about how things appeared to us on other occasions to formulate generalizations about appearances. Instead, we just auto-

---

43. Such a view was advocated by Bertrand Russell (1912, pp. 21-24).

matically make judgments about physical objects as a result of being appeared to in various ways. Nor, it seems, are we doing anything epistemically objectionable by making perceptual judgments in this way. Thus scientific realism imposes unreasonable burdens on rational perceivers.[44]

We turned to induction as a way of accommodating the fact that perceptual knowledge cannot proceed entirely in terms of conclusive reasons. Inductive reasons used to be the only nonconclusive reasons that were considered uncontroversial in contemporary philosophy, so they provided the obvious candidate for the nonconclusive reasons involved in perceptual knowledge. But it now seems that perceptual knowledge cannot be reconstructed in terms of inductive reasons. If perceptual knowledge requires nonconclusive reasons, and those reasons are not inductive reasons, then it is inescapable that there must be noninductive nonconclusive reasons. And once we have made this admission, an obvious hypothesis is that what justifies our ordinary inference from "x looks red to me" to "x is red" is the simple fact that the former is a defeasible reason for the latter. Upon reflection, this seems rather obvious. Normally, we do not hesitate to judge that something is the way it looks to us. We only desist from that inference when we have information that constitutes a defeater. For example, if I see a book and it looks red to me then I will normally judge that it is red. But if I am assured by its author that it is not red, this may give me pause and make me reconsider my judgment. We can account for this by noting that the report of the author constitutes a defeater for the defeasible reason. Alternatively, if I learn that I am viewing the book under red lights, and I know that red lights can make things look red when they are not, that may also make me withhold judgment about the color of the book. This is because it gives me a reason to doubt that it would not look red to me unless it were, and that is a defeater for the defeasible reason.

We propose that this is the only way a foundations theory can handle the problem of perception. The epistemic ascent involved in perceptual knowledge must proceed in terms of noninductive defeasible reason. This constitutes more of a "resolution" of the problem of perception than a solution. It does not solve the problem by reducing it to something deeper, but rather eliminates the problem by claiming that there is nothing

---

44. An objection to scientific realism that may be logically more conclusive emerges from the realization that there is a problem about memory that is analogous to the problem of perception. This problem is discussed at length in the next section. As will become apparent then, if we were to attempt to apply scientific realism to the combination of perceptual and memory knowledge, we would not be able to do so because the "data" available to us at any given time would be severaly impoverished, never consisting of more than about seven items.

deeper and the inferences involved in perceptual knowledge are primitive constituents of our epistemic framework. This contrasts with traditional attempts, which tried to solve the problem by justifying the inferences on the basis of more complicated arguments proceeding in terms of allegedly simpler kinds of reasons.

Once we have acknowledged the existence of a few noninductive defeasible reasons, other candidates will occur quite naturally. This will be illustrated in section four when we consider the role of memory in reasoning.

3.3.5 Defeasible Reasons

We have argued that defeasible reasons play an indispensable role in our ratiocinative framework and have illustrated their role in epistemic ascent through the problem of perception. Without them, knowledge of the world would be impossible. At the very least, both perception and induction proceed in terms of defeasible reasons, and it is arguable that this is true of other areas of knowledge as well.[45] The importance of defeasible reasons is also illustrated by another phenomenon. Gilbert Harman (1973) has called attention to the fact that reasoning can lead not only to the adoption of new beliefs but also to the rejection of old beliefs. If all reasoning proceeded exclusively in terms of conclusive reasons, that would be inexplicable. Conclusive reasons are nondefeasible, so if we are justified in believing P and P gives us a conclusive reason for Q and we come to believe Q on that basis, there is nothing that could rationally make us retract our endorsement of Q except by making us retract our endorsement of P. And if P is also held on the basis of conclusive reasons, the same thing goes for it. Thus if all reasoning were conclusive, nothing could ever rationally commit us to taking anything back (except by retracting some of our initial appearance judgments, which we never or virtually never do). It is the fact that we reason in terms of defeasible reasons, and the latter are defeasible, that explains how reasoning can lead to the rejection of previously held beliefs. Reasoning accomplishes this by producing defeaters for the reasons for which we hold the beliefs. For example, I may justifiably believe that something is red because it looks red to me. I have a justified belief here. But if I then acquire a further belief, viz., that the lighting is peculiar in certain ways, that may justify me in believing a defeater for the original defeasible reason and may have the result that I am no longer justified in believing that the object in question is red. If I am rational, I will then reject that belief.

---

45. For further discussion of this, see Pollock (1974) and (1989), and chapter seven of this book.

4. Reasoning and Memory

On the foundationalist's proposal, epistemic ascent proceeds by reasoning, which leads to the formation of new beliefs on the basis of previously held beliefs. But now a surprising new issue arises. Just what are the beliefs from which we can form new beliefs on the basis of argument?

4.1 Occurrent Thoughts

We have thus far implicitly adopted a kind of "mental blackboard" picture of reasoning according to which (1) we have an array of interconnected beliefs all available for simultaneous inspection and evaluation and (2) arguments are built out of these beliefs and are evaluated by such inspection. That is the picture normally adopted, but it is unrealistic. To see this, let us begin by distinguishing between thoughts and beliefs. At any given time, we are not thinking about most of the things we believe at that time. We all believe that 2+2 = 4, but this is not something that is likely to have "occurred to" the reader in the past five minutes. It is not something that she has actually *thought*. Thoughts, on the other hand, are what we are occurrently thinking. At any given time we are apt to have many beliefs but few thoughts. It is difficult to hold very many thoughts in mind at one time. In particular, we rarely hold an entire argument (even a simple one) in mind at one time. Psychological evidence indicates that people can hold about seven items in mind at one time. There is some evidence that the number may be even smaller for complex items like complicated propositions.[46]

The term "thought" is normally used to refer to either occurrent beliefs or to a more general class of mental events that includes our entertaining ideas without mentally endorsing them. In the latter sense thoughts may include hypotheses, fears, musings, daydreams, and so on. However, we will restrict our use of the term to occurrent beliefs. So thoughts are beliefs, but most beliefs are not thoughts. Given this distinction, which are involved in arguments and in the determination of justification—beliefs in general or just thoughts? Reasoning is an occurrent process, so it might seem that insofar as justification emerges from reasoning it can only be thoughts that enter into considerations of justification. The trouble with this is that we have too few occurrent thoughts at any one time to be able to construct arguments out of them. Although reasoning is an occurrent process, that does not mean that we occurrently hold an entire argument in mind. Rather, we progress through the argument one step

---

46. The classic article on the limits of memory is George Miller (1957). Also see W. Kintsch and J. M. Keenan (1973), and W. Kintsch (1974). For a thorough overview, see the chapters on memory in John Anderson (1995).

at a time, occurrently holding each step in mind as we come to it but not holding the entire argument in mind. Memory plays an indispensable role in such reasoning, in at least two ways. On the one hand, we employ memory to supply us with premises for arguments. These premises will typically be the conclusions of earlier arguments, but we do not have to rehearse those arguments in order to make new use of their conclusions. We also keep track of the course of an argument by relying upon memory to ensure that the first part (which we are no longer holding in mind) went all right and to alert us when there is a step in the argument for which we subsequently acquire a defeater.

4.2 Memory as a Source of Knowledge

How do these observations about the role of memory in reasoning fit into the foundationalist picture of epistemic justification? The foundationalist must say different things about the different roles played by memory in reasoning. Let us begin with what might be called "premise memory". Most of the information at our disposal at any given time is stored in memory and recalled when we need it. What are we to say about the justifiedness of beliefs held on the basis of memory? Reasoning is an occurrent process. It can proceed only in terms of what we occurrently hold in mind. We do not have to hold the entire argument in mind in order for it to justify its conclusion, but we do have to hold each step in mind as we go through it. Thus memory can only contribute premises to an argument insofar as we occurrently remember those premises. Furthermore, we can have varying degrees of difficulty in recalling beliefs that are stored in memory. If we remember something (hold it in memory) but are unable to occurrently recall it just now, then it can play no role in justifying new beliefs. In other words, only occurrent memory can supply premises for arguments.

Granted that only occurrent memory can supply premises, what are we to say about the justification of those premises and the justification of conclusions inferred from those premises? A common view has been that when we hold a belief on the basis of remembering it, what determines whether the latter belief is justified is the argument we instantiated when we first acquired it.[47] On this picture we have an evolving network of arguments that grows longer and more complex over time. Old arguments are extended as we continue to reason from their conclusions, and new arguments are added as we acquire new basic beliefs and reason from them, but the old arguments do not drop out of the picture just because

---

47. See Norman Malcolm (1963), 229-230, and Robert Squires (1969). Of course, we can also come to instantiate new arguments for old beliefs, in which case the source of justification may change, but the view is that that is not what is involved in memory.

we are no longer thinking about them. They continue to represent the justificatory structure underlying our beliefs.

Critics of foundationalism commonly associate this picture with foundationalism,[48] but foundationalists need not adopt such a picture and they would be well advised not to. The difficulty with the picture is that it overlooks some important facts about memory. It has already been noted that we can have varying degrees of difficulty recalling things, and our memory is not infallible. Sometimes we "remember" incorrectly. When that happens, what are we to say about the justifiedness of beliefs inferred from the incorrect memories? We do not automatically regard a person as unjustified in holding a belief just because that belief is inferred from false memories. If he has no reason to suspect that his memory is faulty, we regard his behavior as epistemically beyond reproach. This is true even if he is misremembering. For example, consider a person who has all of his memories altered artificially without his knowing it. Is he then unjustified in everything he believes? Surely not. Recall that when we talk about justification we have in mind the reason-guiding sense of justification. If a person has no reason to be suspicious of his apparent memories, then he is doing the best he can if he simply accepts them. Consequently, he is justified. But if he is misremembering, the belief in question is not one that he previously held or for which he previously had reasons. This seems to indicate that it is the process of remembering itself that confers justification on the use of a memory in a present argument, and not whatever reasons one may or may not have had for that belief originally.

The only way the foundationalist can allow that the process of remembering can confer justification on a belief is by supposing that memory provides us with epistemologically basic beliefs. It is important to realize that *what* is remembered can be a proposition of any sort at all. Sometimes there is a temptation to suppose that we can only remember facts about the past, but memory is just the process of retrieving stored information, and that information can be of any sort. For example, I can remember that 4+7 = 11. This is a timeless truth. I can remember general truths, e.g., that birds fly. And I can even remember facts about the future, such as that there will not be another solar eclipse visible in North America until 2032. By definition, epistemologically basic beliefs comprise a privileged subclass of the set of all possible beliefs, so it cannot be true that the proposition remembered is always epistemologically basic. Rather, memory must operate on analogy with sense perception. Sense perception provides us with beliefs about material objects, but according to foundationalism it does so only indirectly by providing us with beliefs

---

48. See Gilbert Harman (1984).

about appearances from which we can infer beliefs about material objects. Similarly, if we are to accommodate memory within foundationalism, memory must provide us with beliefs about what we "seem to remember" and then we infer the truth of what are ordinarily regarded as memory beliefs from these apparent memories. The viability of such an account turns in part on whether there is such a psychological state of "seeming to remember" that is analogous to being appeared to in some way or other. Some philosophers have denied that there is any such state,[49] but it is not too hard to see that they are wrong. It is possible to hold the same belief on the basis of memory, or perception, or for no reason at all, and when we hold the belief we can tell introspectively which is the case. In other words, we can discriminate between memory beliefs and other beliefs.[50] But to say this is just to say that memory has an introspectively distinguishable mental characteristic. The mental state so characterized is the state of "seeming to remember". This can be made clearer by considering an example. Imagine that you are trying to quote the first line of a poem. It is on the tip of your tongue, but you cannot quite get it. Finally, a friend tires of watching you squirm and tells you the line. This can have two possible effects. It may jog your memory so that the line comes flooding back and you now remember it clearly. Alternatively, it may fail to jog your memory. You believe your friend when he tells you how the lines goes, but you still do not remember it. In either case you come to have the same occurrent belief about the line, but there is a clear introspectible difference between the two cases. The difference is precisely that in the first case you come to be in the state of seeming to remember that the line goes that way, whereas in the second case you have no such recollection. Cases like this show that there is such a psychological state as that of seeming to remember.

Given that there is such a state as seeming to remember, the natural move for the foundationalist is to treat memory as a source of knowledge parallel to sense perception and posit the following "mnemonic" defeasible reason:

"S seems to remember P" is a defeasible reason for S to believe P.

This becomes the foundationalist's explanation for how memory can supply premises for arguments that confer justification on new beliefs. Furthermore, it seems to be the only possible way to integrate premise memory into a foundationalist theory.

---

49. Norman Malcolm (1963), 229-230, and Robert Squires (1969).
50. By "memory beliefs" we mean "putative memory beliefs". We do not mean that we can tell introspectively whether we are correctly remembering what we take ourselves to be remembering.

4.3 Genetic Arguments and Dynamic Arguments

What about the other aspects of memory as it is used in reasoning? We were led to the topic of memory by the observation that the mental blackboard picture of reasoning is wrong. We do not hold an entire argument in mind at one time. Rather, we step through it sequentially, holding no more than a few lines at a time in occurrent thought. Insofar as we have to know that the earlier parts of the argument were all right, we must rely upon memory. It is tempting to try to assimilate this use of memory to premise memory in the following way. Suppose we reason through the complicated argument in Figure 2.3 and on that basis come to believe P_n.

Argument (1)

Figure 2.3

As we occurrently step through the ith line of the argument we may occurrently recall nothing earlier than the i-1st line. At that point, only memory can certify that the earlier parts of the argument were all right. This suggests that the basis upon which we actually come to believe P_i is not argument (1) at all, but rather a much shorter argument whose first premise is supplied by memory; see argument (2), Figure 2.4. Having inferred P_{i-1}, in order to proceed to P_i all we have to do is remember P_{i-1}. Premise memory certifies P_{i-1}, and then we infer P_i from P_{i-1}. This is what justifies us in coming to believe P_i.

Argument (2)

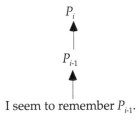

I seem to remember P_{i-1}.

Figure 2.4

But this is puzzling. It seems to indicate that argument (1) is not doing any work. It is not on the basis of that argument that we become justified in holding the individual beliefs comprising it. That argument represents the historical genesis of the beliefs, but in an important sense it does not represent the dynamics of their justification. The latter is represented by lots of little arguments of the form of argument (2). An apparent problem for this view is that we do not regard argument (1) as irrelevant to our justification. If we discover inadequacies in early stages of argument (1) (e.g., if we acquire a defeater for a defeasible reason used early in argument (1)), we take that to make us unjustified in holding the later beliefs in the argument. How can it do that if we do not hold those beliefs on the basis of argument (1)?

Argument (1) and argument (2) are both important in understanding the justification of P_i. We will distinguish between them by calling them the *genetic argument* and the *dynamic argument*, respectively. Note that as we are using the term, dynamic arguments do not always begin with apparent memories. They might begin instead, for instance, with an appearance belief and infer a physical-object belief. The important thing about dynamic arguments is that they represent what we are currently thinking. To use a computer metaphor, the dynamic argument is short and fits into "working memory". We can regard the mental blackboard picture as true of dynamic arguments.

The genetic argument and the dynamic argument are both relevant to justification, but in different ways. The dynamic argument is "positively relevant" in that it tells us what makes us currently justified in believing P_i. The genetic argument is not positively relevant in the same way. If we are no longer able to recall the earlier steps of the genetic argument, it can play no positive role in the justification of our occurrent belief in P_i. On the other hand, the genetic argument is "negatively relevant" to the justification for our occurrently believing P_i, because if (a) we know that the genetic argument underlies our having come to believe P_i and (b) we acquire a defeater for some step of the genetic argument, we regard that as defeating our justification for P_i.

How can we put these observations together into a coherent account of the relationship between reasoning and justification? Earlier it was suggested that justification can be identified with holding a belief on the basis of an undefeated argument. But justification *cannot* be identified with holding a belief on the basis of an undefeated genetic argument. The genetic argument for a belief may stretch back over a period of years as you slowly accumulate the diverse premises. If you can no longer recall the arguments for some of those premises (which is quite likely), then if you presently acquire a defeater for one of those early steps in the argument but do not realize that it is a defeater or do not in any way appreciate its relevance to P_i, we would not regard the acquisition of that defeater as making it unreasonable for you to believe P_i. (Intuitively, this is because you currently believe P_i on the basis of the dynamic argument rather than the genetic argument.) Consequently, it is not a necessary condition for justified belief in P_i that your genetic argument for P_i be undefeated.

Can we instead identify justification with holding a belief on the basis of an undefeated dynamic argument? We can once we recognize that genetic arguments play a role in determining whether a dynamic argument based on memory is undefeated. Such a dynamic argument proceeds in terms of the following mnemonic defeasible reason:

"I seem to remember P" is a defeasible reason for me to believe P.

It is obvious upon reflection that one kind of defeater for this defeasible reason is any reason for thinking that I do not actually remember P. A necessary condition for remembering P is that one originally knew P. If, for instance, my original belief in P was unjustified, but was retained in memory, then even though I now *seem* to remember P, it would be incorrect to describe me now as remembering P. Thus, any reason for thinking that I did not originally know P is also a reason for thinking that I do not remember P now. I did not originally know P if there is a true undefeated defeater for some step of the reasoning (i.e., the genetic argument) underlying my belief in P. Thus the following is a defeater for the mnemonic defeasible reason:

Q is true, and Q is a defeater for some step of my genetic argument for P.[51]

We will call this "the genetic defeater". The concepts of a defeater and a genetic argument seem like technical philosophical concepts not shared by the person on the street. On this ground, it might be doubted that ordinary people actually have beliefs of the form of the genetic defeater. But our contention is that they do. They could not formulate them using this technical terminology, but they could formulate them less clearly by saying things like, "In coming to believe P in the first place I assumed A and concluded B, but because Q is true, I should not have done that." There does not seem to be anything psychologically unrealistic about supposing that people often have thoughts they could formulate in such a way as this, and these are the thoughts expressed more precisely by the above formulation of the genetic defeater.

Our proposal is that, on a foundationalist picture, justification should be identified with holding a belief on the basis of an undefeated dynamic argument. What we mean by this is that the arguments to which we appeal in determining whether a dynamic argument is undefeated must all be in working memory at the same time as the dynamic argument. This makes the requirement of being undefeated a rather minimal one, because we cannot get much into working memory at one time. From one point of view, formulating the requirement in this way seems obviously correct—if we do not occurrently remember an argument for a defeater, we *cannot* take it into account in deciding what to believe, and so we should be deemed epistemically beyond reproach if we ignore it. On the other hand, this makes the requirement that the dynamic argument be undefeated seem so weak as to be virtually useless. It seems we will almost never have any defeating arguments in working memory at the same time as the argument to be defeated, and so a dynamic argument will almost always be undefeated in this sense. We contend, however, that this objection is wrong. As a matter of psychological fact, when we acquire a new belief that constitutes a defeater for the genetic argument, we often remember that it does, and thus we add the dynamic argument in Figure 2.5 to working memory.

---

51. There are other kinds of defeaters as well. The kind of defeater operative in cases where we discover that we are misremembering is "I did not originally believe P".

Argument (3)

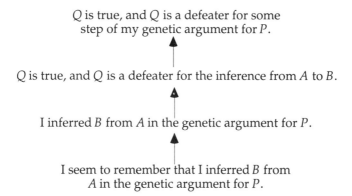

Q is true, and Q is a defeater for some
step of my genetic argument for P.

Q is true, and Q is a defeater for the inference from A to B.

I inferred B from A in the genetic argument for P.

I seem to remember that I inferred B from
A in the genetic argument for P.

Figure 2.5

Thus working memory comes to contain both argument (2) and argument (3), and the latter is a defeating argument for the former. Hence argument (2) is not undefeated.

4.4 Primed Search

According to this account, if we have a defeater for the inference from A to B but we do not remember inferring B from A in the genetic argument for P, then our present belief in P is justified. That seems to be correct. The reason this does not trivialize the requirement that the dynamic argument be undefeated is that we frequently have the requisite memories. Memory allows us to monitor the course of our arguments and alerts us when we subsequently encounter a defeater for an earlier stage of an argument. This is how memory supplies us with genetic defeaters. This aspect of memory is intriguing, partly because it does not fit into naive models of memory that would attempt to reduce all memory to premise memory. In this connection, it might be tempting to suppose that we acquire genetic defeaters by occurrently recalling the earlier parts of the genetic argument and inspecting them to see whether each newly drawn conclusion constitutes a defeater for any of the earlier steps. This would reduce the monitoring function of memory to premise memory. But obviously, we do not really do it that way. To suppose we must is to adopt a simplistic view of memory. This is best illustrated by considering memory searches. When we search our memory for something, we do not have to proceed sequentially through all the beliefs held in memory, calling each to consciousness, inspecting it to see if it is what we are looking for, and if it is not, then rejecting it and going on to the next item. For example, in trying to remember someone's name, despite the fact that this is something that we can voluntarily undertake to do, the process whereby we do it is not a conscious process. We set ourselves to do it, and then we wait a moment and see whether anything emerges into occurrent thought. If we are able to remember the name, the only thing that occurs at a conscious level is the recollection of the name. If we are unable to remember the name, we may feel frustrated and we may continue to "try" to remember it, but nothing happens at the conscious level. The point is that, to use computer jargon, human recollection involves built-in search procedures. If asked to name a famous composer, my memory can find one. It does it by searching my unconscious memory for someone remembered as a famous composer, and because it is searching unconscious memory, this search is not something I do consciously.

Human recollection also involves a somewhat more complicated operation that we might call "primed search". Consider a birdwatcher who has a mental list of rare birds he would like to observe. This need not be a fixed list. Each month he may add new birds to the list when he reads about them in his birdwatcher's magazine, and he strikes items off the list when he observes them. Furthermore, the way in which the list evolves need not consist of his recalling the entire list to mind and then altering it. He can alter such a mental list by adding or deleting items without ever thinking about the list as a whole. Given such a list, when our birdwatcher sees a bird on the list, he immediately recalls that it is one of the listed birds and he may get very excited. The point is that one can prime oneself to be on the lookout for things on such a nonoccurrent mental list. This is an unconscious mental function that humans are capable of performing, and it involves memory in accessing the list itself, but the memory processes involved cannot be reduced to any kind of simple memory of individual facts.

Primed search is what is involved in monitoring our reasoning and being on the lookout for newly inferred defeaters for previous steps of reasoning. We remember what those earlier steps were, although we do not do so occurrently, and we remain on the lookout for defeaters for those earlier steps, and when we encounter such a defeater we then occurrently remember the earlier step and note that we have a defeater. The only conscious output of this primed search consists of occurrently remembering that a certain step occurred in the argument, and we combine that with the observation that a particular newly acquired belief is a defeater for that step. This is how we acquire a genetic defeater for a dynamic argument.

To recapitulate, reflections on the role of memory in reasoning have led us to a radically different picture of the relationship between reasoning and epistemic justification. We have been led to distinguish between the genetic argument for a belief and the dynamic argument, concluding that it is only the latter that is directly relevant to the assessment of the belief as justified or unjustified. The genetic argument is indirectly relevant because genetic defeaters for the dynamic argument appeal to the genetic argument and are supplied by some of the more sophisticated operations of memory. Without these innovations, foundationalism would be much less plausible.

5. Reconsideration of Epistemologically Basic Beliefs

Our purpose thus far has been to sketch as plausible a foundations theory as possible. A foundations theory has two parts: (1) an account of epistemologically basic beliefs, and (2) an account of epistemic ascent. With regard to (1), we have argued that although epistemologically basic beliefs might be incorrigible, that is not a strict requirement of foundations theories. All that is really required is that they be self-justifying, and the weakest kind of self-justification is prima facie justification. With regard to (2), we have maintained that epistemic ascent is a matter of reasoning and must proceed in terms of a combination of prima facie and conclusive reasons. We have also urged that memory plays a much more involved role in reasoning than has generally been recognized.

We believe that the picture we have drawn of reasoning is basically correct and we will employ parts of this picture in our own positive view, but there are fundamental problems for the foundationalist claims about epistemologically basic beliefs and their role in knowledge. In fact, we contend that all foundations theories are false, and that where they go wrong is in their claims about epistemologically basic beliefs. Recall that foundations theories are doxastic theories. Doxastic theories assume that the justifiability of a belief is a function of one's overall doxastic state. The motivation for this assumption is the idea that in deciding what to believe you can only take account of something insofar as you have beliefs about it. Thus only beliefs can contribute to the determination of what you can be justified in believing. There are two kinds of doxastic theories—foundations theories and coherence theories. What distinguishes a foundations theory from a coherence theory is that according to a foundations theory there is a privileged subclass of beliefs (the epistemologically basic beliefs) that have two properties: (1) they are self-justifying; and (2) all other beliefs are justified ultimately by appeal to epistemologically basic beliefs. This was originally motivated by the observation that all justified belief derives ultimately from the evidence of our senses, and the doxastic assumption that that evidence must come to us in the form of beliefs. If this is right and if we are ever to be able to get started in the acquisition of such sensory beliefs, it cannot be required that the justification of the sensory beliefs involves justificatory appeal to other beliefs, so the sensory beliefs must be self-justifying. Further substance was lent to this picture by the observation that we seem to treat sensory beliefs as incorrigibly justified, and that requires them to be incorrigible.

5.1 Incorrigible Appearance Beliefs

If appearance beliefs are incorrigible, that suits them for service as epistemologically basic beliefs, but are they incorrigible? There is no consensus on this issue. There is something intuitively appealing about the thesis of incorrigibility, but there is also something perplexing about the idea that *any* belief could be incorrigible. What could make a belief incorrigible? That requires there to be some kind of strong logical relationship between having the belief and what the belief is about. For ordinary beliefs, holding the belief provides no logical guarantee that it is true, but there might be such a relationship in the case of appearance beliefs. It has occasionally been suggested that the requisite connection is a simple one—there is no difference between the state of being appeared to in a certain way and the state of thinking you are appeared to in that way. According to this view, these are two ways of describing one and the same mental state. We might call this *the identity thesis*. But note that if it is to explain both parts of the definition of incorrigibility, the identity thesis might include a second identity alleging that the state of not being appeared to in a certain way is the same state as that of thinking you are not appeared to in that way. Unfortunately, this putative identity is totally implausible. When, for example, you are not appeared to redly, you rarely have any thoughts about the matter, and so you can be in the state of not being appeared to redly without thinking that you are, and hence these must be two different states.

The identity thesis is false, but its failure suggests a more conservative hypothesis about the origin of incorrigibility. It would suffice for incorrigibility if the state of thinking that one is appeared to in a certain way merely *contained* the state of being appeared to in that way as part of it, and similarly for thinking that one is not appeared to in that way. Then one could not think one was appeared to in a certain way without being appeared to in that way. This *containment thesis* is not implausible. One way of defending the containment thesis is to urge that in order for one to think he is appeared to in a certain way he must focus his attention mentally on that way of being appeared to, and when he does that he *is* appeared to in that way. In other words, in the kind of appearance belief that is incorrigible one is appeared to in a way that involves a kind of mental demonstrative reference wherein he focuses his attention on some feature of his sensory experience and then thinks to himself, "I am appeared to *that* way". A necessary condition for one to be able to have the latter belief is that his sensory experience exhibits the feature to which he is attending, so it seems his belief cannot be mistaken. He cannot think he is appeared to in that way without being appeared to in that way, and hence the belief is incorrigible.

At one time one of us (Pollock) found this reasoning convincing,[52] but

---

52. Such an argument was given in Pollock (1974). Moritz Schlick (1959) made remarks that at least suggest this argument.

now it seems to us to be wrong. One difficulty is that the most it establishes is incorrigibility in the sense of the original, provisional, definition. It purports to show that you are appeared to in a certain way if you think you are, but it is of no help in showing that you are *not* appeared to in that way if you think you are not, and both of these are required for incorrigibility.[53]

We doubt that the argument is even successful in showing that you cannot be wrong in thinking you are appeared to in a certain way. The argument appeals to what we have called "mental demonstrative reference". In ordinary demonstrative reference, you can purport to refer to something that is not there, for example, Macbeth can purport to refer demonstratively to the putative dagger. Why should the fact that you purport to refer demonstratively to a way of being appeared to guarantee that, unlike the ordinary case of demonstrative reference, you really are appeared to in that way? There is an answer to this question, but we contend that it is mistaken. The answer is that your purporting to refer demonstratively to a way of being appeared to does guarantee something—namely, that you have a mental representation of that way of being appeared to. And it is apt to seem that the mental representation of the way of being appeared to is the same thing as the way of being appeared to—the way of being appeared to is self-representing. This, it may be claimed, is why you cannot be wrong in thinking you are appeared to in that way. But are ways of being appeared to self-representing? That is, can you think of a way of being appeared to simply by *being* appeared to in that way? It seems not, for the simple reason that you are normally unaware of being appeared to in a particular way when you are appeared to in that way. In such a case you are appeared to in that way without having the thought that you are, so being appeared to in that way cannot constitute thinking about that way of being appeared to. What this shows is that in ordinary perception you do not think about the way of being appeared to just by being appeared to in that way—you do it by being appeared to in that way and *thinking about it* in the way you can introspectively think about ways of being appeared to when you are appeared to in those ways. But this is now trivial. There is nothing self-representing here.

We have surveyed the major arguments that have actually been deployed in an attempt to show appearance beliefs to be incorrigible. The arguments do not withstand scrutiny, but the lack of good arguments does not establish that such beliefs are not incorrigible. They might be incorrigible anyway. Most philosophers who have claimed that appearance beliefs are incorrigible have done so without argument, on the grounds that it just seems right that appearance beliefs cannot be mistaken.

---

53. We owe this point to James van Cleve, in conversation.

5.2 Problems with Incorrigible Appearance Beliefs

Despite its popularity, the incorrigibility of appearance beliefs has also been denied rather frequently. There are two arguments that are sometimes given against incorrigibility. The first is the "super electroencephalograph argument".[54] Suppose brain physiologists located that part of the brain whose neural activation constitutes a person's being appeared to redly, and they created a super electroencephalograph that monitored that region of the brain. Suppose then that one of these brain physiologists thought he was being appeared to redly, but his colleagues were monitoring his brain and assured him that he was not. Assuming that the evidence for the reliability of the super electroencephalograph was good enough, would it not be unreasonable for him to insist that he really was being appeared to redly and the super electroencephalograph was wrong? This is intended to show that the belief that one is appeared to redly is not incorrigibly justified and hence not incorrigible. But the argument is question-begging. It would only be unreasonable for the scientist to continue to insist that he was appeared to redly if such beliefs were not incorrigibly justified, and that is the very question at issue. *Perhaps* such insistence would be unreasonable, but we need more of an argument to establish this.

The second argument against incorrigibility can be traced to Wilfrid Sellars (1963). According to Sellars, to say that a person is appeared to redly is to say that he is appeared to in the way one is normally appeared to in the presence of red objects. But how red objects appear to a person is a contingent matter that can only be discovered inductively. Thus, rather than being incorrigible, the belief that one is appeared to redly is based upon induction and prior knowledge of physical objects. The standard response to this argument invokes Chisholm's distinction between comparative and noncomparative appearance judgments that we discussed earlier. Recall that comparative appearance judgments classify the way one is appeared to as being the way one is normally appeared to, while noncomparative appearance judgments simply tell how one is appeared to. Sellars' argument against incorrigibility only applies to comparative appearance judgments, and the obvious response is that the foundationalist only means to be claiming that noncomparative appearance judgments are epistemologically basic.

Armstrong and Sellars' arguments against the incorrigibility of sensory beliefs are not decisive. In order to secure a more effective argument, however, consider shadows on snow. Because shadows on white surfaces are normally grey, most people think that shadows on snow are grey. But a discovery made fairly early by every landscape painter is that they are actually blue.[55] A person having the general belief that shadows on

---

54. This argument is, apparently, due to David Armstrong.

55. This is because the light reflected from snow is not actually white—it is blue. After all, snow is water. It looks white to us because the intensity of the reflected

snow are grey may, when queried about how a particular snow-shadow looks to him, reply that it looks grey, without paying any serious attention to his percept. His belief about how it looks is based upon his general belief rather than inspection of his percept, and is accordingly wrong. This shows that the belief is not incorrigible.

5.3 Prima Facie Justified Beliefs

Attacks on foundationalism have tended to focus on incorrigibility,[56] but foundationalism does not require incorrigibility. What foundationalism requires is self-justification, which is weaker than incorrigibility. The attraction of incorrigibility is that it offers an explanation for why certain kinds of beliefs might be self-justifying. But even if we were to decide that epistemologically basic beliefs are not incorrigible, it might still be possible to hold that they are self-justifying for some other reason.

It can be argued that a proposition and its negation could not both be incorrigibly justified without the proposition being incorrigible: suppose P is not incorrigible for S. This means that it is either possible for S to believe P and be wrong or possible for S to believe $\sim P$ and be wrong. Suppose it is the former (the latter case being analogous). S could then discover inductively that there are certain kinds of circumstances under which he has a propensity to believe P and be wrong (just as we discover that under certain kinds of circumstances objects tend not to be the colors they look to us). Then if S believes P but knows he is in such circumstances, it seems he is unjustified in believing P, and hence that belief is not incorrigibly justified. Note, however, that the belief might still be prima facie justified. What makes justification fail in this example is that S has a reason for thinking he should not believe P, and it was precisely to accommodate such "defeaters" that we were led to the concept of prima facie justification.

If we make the assumptions that the negation of an epistemologically basic proposition is itself epistemologically basic, and all epistemologically basic propositions have the same fundamental epistemic status, then it follows that if epistemologically basic beliefs are not incorrigible then they must be only prima facie justified. Thus a possible, and somewhat attractive, version of foundationalism posits the existence of prima facie justified epistemologically basic beliefs. Such a theory should be more appealing to those who are suspicious of incorrigibility. The theory does have a weakness, however, that is not present in more traditional incorrigibility theories. Incorrigibility provides an *explanation* for why epistemologically basic beliefs are self-justifying, but if we take epistemologically

---

light washes out the color.

56. Such attacks have been pressed by both Keith Lehrer (1974) and Richard Rorty (1979).

basic beliefs to be merely prima facie justified then we are just positing self-justification without any explanation. Furthermore, prima facie justification can seem puzzling. How could any belief be prima facie justified? What could confer such a status on a belief? That seems mysterious and in need of explanation.

5.4 Problems with Prima Facie Justified Beliefs

The example of shadows on snow can also be used to argue that the belief that the snow shadow looks grey is not epistemologically basic by virtue of being prima facie justified. Suppose again that you have the general belief (arrived at inductively) that shadows on white surfaces are grey, and the belief about how a particular snow shadow looks is based upon a general belief about how shadows look on white surfaces rather than on inspection of your percept. Accordingly, the belief is wrong. Suppose further that your inductive evidence is faulty and you are unjustified in believing that shadows on white surfaces look grey. Then you are unjustified in believing that the snow shadow looks grey. If the belief is prima facie justified, you can only be unjustified in holding it if you have some reason for thinking you should not hold it. Do you have any reason for thinking that you should not believe that the snow shadow looks grey? The answer to this question is a bit complicated. There is some temptation to claim that you automatically have such a reason with regard to any belief that you hold for bad reasons. This is an epistemic principle that will be explored below in more detail. It has the consequence that this example does not show that appearance beliefs are not prima facie justified, but this is only because it makes *all* beliefs prima facie justified. To see this, consider any belief *P* and suppose you are unjustified in believing *P*. Then you do not believe *P* for a good reason, and so by the proposed principle it follows that you have a reason for thinking you should not believe *P*. Thus you can only be unjustified in believing *P* if you have a reason for thinking you should not believe *P*, and hence belief in *P* is prima facie justified. And this argument works for *any P*. Thus, although this epistemic principle would make the belief that you the snow shadow looks grey prima facie justified, it would not make it epistemologically basic. There would be no epistemologically basic beliefs if this principle were true. The result would be a coherence theory rather than a foundations theory, because an essential claim of a foundations theory is that the epistemologically basic beliefs form a privileged *subset* of beliefs on the basis of which other beliefs are justified. So a foundations theorist cannot defend his theory against the snow shadow example by endorsing this epistemic principle. It seems to follow that the belief that the snow shadow looks grey cannot be prima facie justified.

There is a response to this counterexample that has considerable intuitive pull, at least initially. This is to agree that not all beliefs about how things appear to us are prima facie justified, but those *based upon actually being appeared to in that way* are. Taken literally, this makes no sense. Prima facie justification is a logical property of propositions. A proposition cannot have such a property at one time and fail to have it at another. But the claim actually being made here is presumably a different one, viz., that when we are appeared to in a certain way, that in and of itself can make us at least defeasibly justified in believing that we are appeared to in that way. The deeper point of the shadow example is that there is no guarantee within a doxastic foundationalist framework that a person will be consulting her sensory percept when she makes a judgment about how things look. If a theory demanding that cognizers inspect their percepts when reasoning from basic beliefs could be defended, then appearance beliefs might be prima facie justified. But it is clear that no such necessity obtains for beliefs about percepts. Later in the book we will endorse a theory providing such a foundation for epistemic justification, but notice that such a theory is no longer a doxastic theory. The justifiedness of beliefs is no longer determined exclusively by *what* we believe. What percepts we have is also relevant. Thus this is not a way of saving doxastic foundationalism.

5.5 Problems with Foundationalism

We have cast doubt on the possibility of incorrigible or prima facie beliefs within a traditional foundationalist framework. Foundations theories are also subject to a different kind of objection that is ultimately more illuminating. Foundations theories are motivated by the idea that all justified belief derives ultimately from the evidence of our senses, and that evidence comes to us in the form of beliefs. We cannot fault the first part of this motivation, but we think it is a mistake to suppose that the evidence of our senses comes to us in the form of beliefs. We rarely have any beliefs at all about how things appear to us. In perception, the beliefs we form are almost invariably beliefs about the objective properties of physical objects—not about how things appear to us. If only the latter are candidates for being epistemologically basic, then it follows that perception does not usually provide us with epistemologically basic beliefs and hence perceptual knowledge does not derive from epistemologically basic beliefs in the way envisaged by foundations theories. This objection to foundations theories seems to me to be decisive, but let us examine its details a bit more closely. We are urging (1) that a person rarely has beliefs about how things appear, but (2) only beliefs about how things appear are plausible candidates for being the epistemologically basic beliefs underlying perceptual knowledge, so (3) perceptual knowledge is not based upon epistemologically basic beliefs and hence foundationalism is wrong. The foundationalist must deny either (1) or (2).

It is undeniable that we are rarely *aware* of having thoughts about how things appear to us. When I walk into a room I think things like "That is Jones standing over there" and "The couch has been reupholstered in a bright red fabric". I am not aware of thinking things like "A bright red trapezoidal shape appears in the upper right-hand corner of my visual field". I *can* think things like the latter, but that normally involves a conscious shift of attention and a reorientation in my thinking. It does not seem that such thoughts normally occur of their own accord. The only way to deny this is to insist that we have such thoughts but do not normally attend to them and hence are unaware of having them.

If it is granted that we are normally unaware of having thoughts about how things appear to us, we see no reason to believe that we normally have such thoughts. The burden of proof is on the foundations theorist to establish the existence of these thoughts. But we can go further and produce a reason for thinking that people do not ordinarily have such thoughts. If we were only unaware of having such thoughts because we do not attend to them, then all we would have to do to become aware of them is to direct our attention to them. But that seems wrong. It is not that easy to become aware of how things appear to you. It *is* fairly easy to turn your attention inwards and focus it on your sensory experience, but that does not automatically either generate or reveal beliefs about that sensory experience. To explain what we mean, consider the distinction between sensing something and forming beliefs about it. Suppose, for example, that I happen upon an abstract painting that I find very attractive. I may spend some time gazing at it, "drinking it in" perceptually. I am sensing the patterns in the painting, but if they are quite abstract and irregular so that they are not easily categorized, I may not be forming any beliefs about those patterns. This illustrates that consciously sensing something does not *automatically* involve forming beliefs about it (although we should think that in *most* cases of consciously sensing something we do form beliefs about it). Applying this to our ability to turn our attention inward and focus on our sensory experience, what that amounts to is the ability to introspectively sense our sensory experience, but that does not automatically involve our having beliefs about that sensory experience. Psychologically, our perceptual beliefs about the objects in our immediate environment are caused by our having sensory experiences of certain types, and by introspection we can readily become aware of having the sensory experiences, but that does not mean that we also have the beliefs that the sensory experiences are *of* the requisite types. Very often we are not even aware of what the requisite types are. For instance, when face to face with a friend I form the belief that he is before me and I form that belief because I have a certain type of sensory experience, but I have no idea precisely what type of sensory experience it is that leads me to have the belief that my friend is before me. There is a very complicated abstract pattern whose exemplification by my sensory experience causes me to have the belief, but I have no clear idea what that abstract pattern is and accordingly it seems I have no belief to the effect that my sensory experience exemplifies that pattern. Face recognition is an obvious case in which we do not know what patterns are responsible for recognition, but it takes little reflection to realize that the same phenomenon occurs in recognizing something as a vase or a tree or just about anything else. With considerable training one can become aware of some of these patterns. Acquiring such awareness is an important part of becoming a painter, but it is a difficult process. Think what a remarkable discovery it was of the pointillists that the appearance of an object can be reproduced with discontinuous dots of color.

We take all of this to indicate that we do not have the beliefs about appearances that foundations theories would require us to have. We are not ordinarily aware of even *having* sensory experiences. By shifting our attention we can make ourselves aware of our sensory experiences, but that need not involve having beliefs about them and hence it gives us no reason to think that in more ordinary cases we have such beliefs but they are unconscious. Furthermore, the painter examples show that insofar as we do have beliefs about our sensory experience they are often wrong. This is related to the earlier argument we gave to show that such beliefs are not incorrigible. When looking at a snow-covered field I may think that the shadows in the snow look grey to me when they actually look blue to me. This is made possible by the fact that I have the mistaken general belief that snow looks white and shadows in snow look grey, and so without really attending to my current sensory experiences I jump to the conclusion that the shadows look grey in this case.

It must be concluded that beliefs about how things appear to us are ill-suited to play the role demanded of them by foundations theories. We rarely have *any* such beliefs. We most definitely do not have *enough* beliefs of this sort to base all our other beliefs upon them. Furthermore, such beliefs would not provide a firm foundation because they are not self-justifying. More accurately, they are not incorrigible, and although we have not ruled out the possibility that they might be prima facie justified, they can only be prima facie justified if *all* beliefs are prima facie justified. In the latter case all beliefs have the same epistemological status, and hence there is no privileged class of epistemologically basic beliefs and so foundations theories fail anyway.

A related point must be made about memory. We have urged that memory plays a very important role in reasoning, and we attempted to reconstruct this within a foundations theory by endorsing the following mnemonic defeasible reason:

"I seem to remember P" is a defeasible reason for me to believe P.

The trouble is, we do not have beliefs about seeming to remember ("memory experiences") any more often than we have beliefs about sensory experiences. In normal cases of remembering, where we have no reason to doubt our memory, we just remember. The only thought involved in such remembering is what is remembered. We have no thoughts *about* remembering. We usually only have thoughts about seeming to remember when doubt is cast upon the veridicality of our memory. It follows that we cannot account for the justifiedness of memory beliefs by appealing to the mnemonic defeasible reason. It also follows that we cannot accurately reconstruct the role of memory in reasoning by appealing to that defeasible reason. The foundationalist account of memory is no more defensible than the foundationalist account of perception.

We conclude that foundations theories cannot successfully claim that we have the beliefs about appearances and memory experiences that we would be required to have in order to base all knowledge upon epistemologically basic beliefs of these sorts. A foundations theory can only be defended by adopting a different view of what beliefs are epistemologically basic. Recall again that the idea behind foundations theories is that all justified belief derives ultimately from the evidence of our senses, and that evidence comes to us in the form of beliefs. We must either reject this basic idea or adopt a different view of what beliefs formulate the evidence of our senses. In ordinary perception the only beliefs we seem to have are beliefs directly about the objective properties of physical objects. These are beliefs about how things are—not about how things appear to us. Perhaps then we should take these physical-object beliefs to be epistemologically basic. The problem with this proposal is that epistemologically basic beliefs must be self-justifying if they are to provide a foundation for the justification of other beliefs. Physical-object beliefs are clearly not incorrigible, so the only way for them to be self-justifying is for them to be prima facie justified. Consider an ordinary physical-object belief, for example, the belief that my daughter is wearing a coat. I frequently hold such a belief on the basis of perception, and in that case it is usually justified. But suppose now that my daughter has gone to a football game, the evening has turned cold, and I am worried about whether she took a coat. I may think to myself, "Oh, I am sure she is wearing a coat". But then on reflection I may decide that I have no reason to believe that—my initial belief was just a matter of wishful thinking. Prior to deciding that I held the belief on the basis of wishful thinking, was the belief justified? It certainly does not seem so. Even though I had not yet decided that I held the belief on the basis of wishful thinking, that was my basis, and that seems sufficient to make the belief unjustified. Furthermore, not yet having decided that I held the belief on the basis of wishful thinking, it does not seem that I had any reason for thinking I should not hold the belief. So it is apparently possible to hold such a belief unjustifiably without having any reason for thinking that one should not hold the belief, and hence it is not prima facie justified. Consequently, such physical-object beliefs cannot be epistemologically basic.

The above argument is predicated on the assumption that if we hold a belief for bad reasons but have not yet decided that we hold the belief for bad reasons, then we do not have a reason for thinking we should not hold the belief. Suppose that is denied and it is insisted instead that whenever we hold a belief for no reason or for bad reasons then we automatically have a reason for thinking we should not hold the belief. This is the epistemic principle discussed above. Endorsing this principle would vitiate the above argument and make my belief that my daughter is wearing a coat prima facie justified. But as we argued above, this is only because it would make all beliefs prima facie justified. Thus, once again, what is generated is not a foundations theory, but a coherence theory. So this is not a way of defending foundations theories.

This argument against foundations theories is quite general. What we are finding is that all beliefs (even beliefs about appearances) can be held for bad reasons. Depending upon what we say about what counts as a reason for thinking we should not hold a belief, this either has the consequence that no belief is prima facie justified or that all beliefs are prima facie justified. In either case, foundationalism fails, because foundationalism requires there to be a privileged subclass of beliefs that are prima facie justified and on the basis of which all other beliefs are justified. We regard this as a decisive refutation of foundationalism.

If foundationalism fails, what should we erect in its place? We have two options. We can retain the doxastic assumption and adopt a coherence theory, or we can reject the doxastic assumption and adopt a nondoxastic theory. Our ultimate proposal will be that we follow the latter course, but because the doxastic assumption has so much intuitive appeal we will first explore the possibility of adopting a coherence theory. That is the topic of the next chapter.

3
COHERENCE THEORIES

1. Motivation

Doxastic theories begin by assuming that all that can enter into the determination of whether a belief is justifiable is what other beliefs one holds. Foundations theories give a subclass of beliefs—the epistemologically basic beliefs—a privileged role in this determination. We have argued that there is no way to defend epistemologically basic beliefs. The first alternative we will investigate is to retain the doxastic assumption and see if an acceptable theory without basic beliefs can be constructed. A coherence theory is any doxastic theory denying that there is such an epistemologically privileged subclass of beliefs. Coherence theories insist that all beliefs have the same fundamental epistemic status, and the justifiability of a belief is determined jointly by all of one's beliefs taken together. This has typically been expressed by saying that what determines whether a belief is justified is how it "coheres" with the set of all your beliefs. To generate an actual theory of this sort, we must give a precise account of the coherence relation. There is a broad array of possible coherence theories resulting from different ways of analyzing the coherence relation. But they all turn upon the fundamental idea that it is a belief's relationship to all other beliefs, and not just to a privileged subclass of beliefs, that determines whether it is justified.

Coherence theories as a group can be motivated by several different kinds of considerations. The most straightforward is that some coherence theory must be true because all foundations theories are false. This turns upon the doxastic assumption, which we will eventually reconsider, but most philosophers have been inclined to grant the doxastic assumption.

There is a second motivation that has been equally prominent in persuading people of the correctness of coherence theories. This motivation is quite separate from the technical failure of foundations theories, and is well expressed by the Neurath metaphor cited in chapter one: "We are like sailors who must rebuild their ship upon the open sea." At any given time we have a large stock of beliefs—our "doxastic system"—and among these are beliefs telling us how to go about modifying this very stock of beliefs, adding new ones and rejecting old ones. We cannot forsake all of our beliefs and start over again, because then we would not know how to start. To proceed rationally, we must have beliefs that direct our way of proceeding. But these procedural beliefs are not sacrosanct either, because if our beliefs about how to proceed conflict too drastically with the bulk of the rest of our beliefs, we will take that to refute the procedural beliefs. As was emphasized in chapter one, the validity of an argument does not determine whether the conclusion should be accepted or the premises rejected. All the validity of the argument determines is that we should not simultaneously accept the premises and the denial of the conclusion. Which of these we reject must be determined by our relative confidence in them. We should reject that of which we are least confident. To this we can add that another option is to reject the validity of the argument. The apparent conflict between premises of which we are confident and a conclusion we are convinced to be false may persuade us that our beliefs about valid inference are in error. The latter are just further beliefs in our stock of beliefs. Our general strategy must be to attempt to render the entire set of beliefs internally consistent by using the procedural beliefs in the set to guide us, the latter also being subject to revision. So we start with the set of beliefs we already have and use its members to guide its own overhaul.

The Neurath picture has played a prominent role in motivating coherence theories,[57] but we think that the philosophical credentials of coherence theories are best disengaged from it. If we had not already rejected foundations theories on the grounds that there are no epistemologically basic beliefs, the Neurath picture would not provide differential support for coherence theories. There are two things wrong with it. Part of the picture turns on a mistake, and the part of the picture that is correct can be accommodated by a foundations theory. The part that is mistaken is the supposition that belief change is always guided by beliefs. It is not. For example, if I infer Q from P and $(P \rightarrow Q)$, this is not because I have the *belief* that Q follows from P and $(P \rightarrow Q)$. Of course, having been trained in logic I do have that belief, but there was a time when I had no such beliefs (I had not even thought about the matter), and most nonphilosophers probably have very few such beliefs. We reason in accordance with rules like *modus ponens*, but to reason in accordance with them and to be guided by them does not require that we be self-reflective about our reasoning to the extent of having beliefs about what rules we use. Furthermore, complicated kinds of reasoning often proceed in terms of rules that even philosophers cannot articulate. A clear example of this is induction. The "new riddle of induction" is the problem of giving a precise account of rules of inductive inference, and it was illustrated in chapter one just how difficult a problem that is. We all reason inductively, but not even experts have been able to formulate rules of inductive inference that systematize our reasoning in all respects. Obviously, we do not guide our inductive reasoning by *beliefs* about correct rules of inductive reasoning. Consequently, it is a mistake to suppose that in modifying our total stock of beliefs we appeal to beliefs in that stock pertaining to how to reason. That may go on to some extent, when we employ learned patterns of reasoning, but most native reasoning is medi-

---

57. See, for example, W. V. Quine and Joseph Ullian (1978), Keith Lehrer (1974), and Gilbert Harman (1973) and (1984).

ated by rules we cannot easily articulate and regarding which we may have no beliefs at all.

Perhaps it should be regarded as an inessential part of the Neurath picture that we have beliefs about how to change our beliefs. A coherence theory could still be defended merely by insisting that in changing our beliefs we reason from the beliefs we already have, and when beliefs conflict all we can do is weigh the conflicting beliefs against one another in terms of our relative confidence in them. This is partly right, but it constitutes an incomplete account of belief change. We do often acquire new beliefs and reject old beliefs by making inferences from the beliefs we already hold. But a source of new beliefs that is at least equally important is perception. We acquire beliefs about our physical environment by perceiving it. Such beliefs are not inferred from other beliefs. Perception is not inference from previously held beliefs, and it is not mediated by beliefs about how to acquire perceptual beliefs. Perception and inference are alike in that both processes go on "under the surface", unmediated by procedural beliefs concerning how to do it. Consequently, no correct account of belief change can appeal exclusively to the beliefs we already have. It must also allow for the introduction into our doxastic system of new perceptual beliefs.

The accommodation of perception is a major problem for any doxastic theory. The most natural view of epistemic justification is the doxastic assumption according to which the justifiability of a belief is a function exclusively of what one believes. Foundations theories try to accommodate the doxastic assumption by supposing that perception issues in beliefs about sensory experience. By supposing that those beliefs are self-justifying, the foundations theorist avoids the question of how they are to be justified, and then tries to reconstruct perceptual knowledge of physical objects in terms of perfectly ordinary belief-to-belief reasoning from the basic beliefs about sensory experience. The difficulty, of course, is that perception does not usually issue in beliefs about sensory experience, and hence perceptual knowledge cannot be regarded as the result of reasoning from basic beliefs. The difficulty for coherence theories, on the other hand, is that it is not clear what alternatives exist that are compatible with the doxastic assumption. How else can we get perceptual beliefs into our doxastic system? We know that foundationalism is false, but that does not help us very much in determining the true account of perceptual knowledge.

Leaving aside the problems of inference and perception, there is a residual core in the Neurath picture that seems unmistakably correct, and it is really this core that made the whole picture seem right in the first place. The core idea is that beliefs are innocent until proven guilty. When we start with a stock of beliefs, they tend to become disengaged from the reasons for which we originally came to hold them. We tend not to remember our reasons—just our conclusions. If we no longer remember our reasons for a belief, then it seems that the credentials of the belief no longer depend upon those reasons. Finding something wrong with the reasons cannot discredit the belief if we have no idea that the belief was originally derived from those reasons. This might seem perplexing. It might seem that we *should* keep track of our reasons. But it is an undeniable matter of fact that we do not. For example, we all believe that Columbus landed in America in 1492, but how many of us have any idea what our original reason was for believing that? We can guess that we learned it from our parents or from a teacher, but we do not really know and we certainly do not know the details. Furthermore, there is a simple explanation for why human beings are not built in such a way that they do habitually remember their reasons. We are information processors with a limited capacity for storage. If we had to remember our reasons in addition to our conclusions, that would clutter up our memory and overstress our limited storage capacity.[58]

The observation that beliefs become disengaged from their reasons and acquire a presumptive epistemic status of their own lends substance to the picture of a web of interrelated and interacting beliefs with epistemic justification consisting of a belief's having a secure niche in this web. We still have to get perception into the picture somehow, and we have to explain how reasoning works in playing some parts of the web off against other parts, but the general result is a structure of the sort that coherence theorists find intuitively appealing. What is often overlooked, however, is that this structure is entirely compatible with a foundations theory. When beliefs become disengaged from their reasons they are remembered. That is what memory is. And any foundations theory is going to contain an account of memory. The theory we sketched in chapter two treats memory as analogous to sense perception and takes seeming-to-remember as providing a defeasible reason for what is remembered. When beliefs become disengaged from their reasons and enshrined in memory we become able to recall them, and when we recall them we believe them on the basis of seeming to remember them. Thus memory itself gives the beliefs an evidential status independent of their original reasons, and we should automatically continue to hold such beliefs unless we acquire a defeater for the mnemonic defeasible reason. When several such memory beliefs come in conflict with each other, they constitute defeaters for each other, and in deciding which to reject we have to weigh the strengths of our various mnemonic reasons. That is the same thing as asking how confident we are of the various beliefs. So the correct part of the Neurath picture is readily accommodated by a foundations theory and constitutes no argument for a coherence theory.[59] Of course, the foundations theory that accommodates it has already been discredited on the grounds that in remembering something we do not

---

58. Gilbert Harman (1984) makes this point.

59. Susan Haack (1993) offers indirect evidence for this in her attempt to craft a hybrid account of justification that she calls *founderentism*. What Haack's theory shows, however, is that one can retain a fundamentally foundational structure while acknowledging the core insight of coherentism.

usually have beliefs about seeming to remember, but this is just an instance of the general objection that we rarely have beliefs of the sort styled as epistemologically basic. So it is not the Neurath picture that should lead us to reject foundations theories, but rather the lack of epistemologically basic beliefs. This observation is important because it shows where the real disagreement between coherence theories and foundations theories lies. They do not disagree about the defensible parts of the Neurath picture. What they disagree about is whether we have epistemologically basic beliefs, and in this we think one must side with the coherence theories.

Because we do not have appropriate epistemologically basic beliefs, all foundations theories are false, but this is not yet to say which theory is true. If the doxastic assumption can be maintained then it will follow that some coherence theory is true. But another option is to deny the doxastic assumption, thus rejecting both foundations theories and coherence theories. Because it is their inability to handle perceptual input in terms of beliefs that leads to the downfall of foundations theories, one might suspect that the real culprit is the doxastic assumption, and that will eventually be our conclusion. But first, let us survey the various possible kinds of coherence theories to see if they can solve the problem.

2. A Taxonomy of Coherence Theories

The essential feature of a coherence theory is that it is a doxastic theory that assigns the same inherent epistemic status to all beliefs. Insofar as we can demand reasons for holding one belief, we can demand reasons for holding any belief; and insofar as we can be justified in holding one belief without having a reason for doing so, we can be justified in holding any belief without having a reason for doing so.

We can classify coherence theories in two different ways—in terms of the role they assign to the reasons, and in terms of the nature of those reasons.

2.1 Positive and Negative Coherence Theories

Some coherence theories take all beliefs to be prima facie justified. According to these theories, if one holds a belief, one is automatically justified in doing so unless she has a reason for thinking she should not. All beliefs are "innocent until proven guilty". This is the view expressed by the Neurath metaphor. According to theories of this sort, reasons function in a negative way, leading us to reject beliefs but not being required for the justified acquisition of belief. We call these *negative coherence theories*. Positive coherence theories, on the other hand, demand positive support for all beliefs. Positive coherence theories require the believer to actually have reasons for holding each of his beliefs. Different positive coherence theories are generated by giving different accounts of reasons.

The motivations for positive and negative coherence theories are different. Negative coherence theories are motivated by something like the Neurath metaphor. Positive coherence theories are motivated more directly by the failure of foundations theories to find epistemologically basic beliefs. Defenders of positive coherence theories share the intuition of the foundations theorist that justified belief requires reasons, but go on to conclude that because there are no epistemologically basic beliefs, when we trace out the reasons for a belief (and the reasons for the reasons, etc.), the tracing can never terminate with epistemologically basic beliefs requiring no further reasons. Either the tracing of reasons must be allowed to go on indefinitely, or else the nature of reasons must be radically different from what the foundations theorist envisages. These two options are reflected by a second distinction yielding a different classification of coherence theories.

2.2 Linear and Holistic Coherence Theories

We have talked about the role of reasons in coherence theories, but for some coherence theories that is a somewhat misleading way of talking. Consider positive coherence theories. Some positive coherence theories embrace essentially the same view of reasons and reasoning as a foundations theory. On this view, P is a reason for S to believe Q by virtue of some relation holding specifically between P and Q. A reason for a belief is either another individual belief or a small set of beliefs,[60] and is not automatically the set of *all* one's beliefs. On this view of reasons it always makes sense to inquire about our reasons for our reasons, and so on. A positive coherence theory adopting this view of reasons will be called *linear*. On a linear theory, if we trace the reasons for a belief, and the reasons for the reasons, and so on, we can never reach a stopping point. If there were a stopping point, it would have to consist of epistemologically basic beliefs, and coherence theories deny that there are epistemologically basic beliefs. There are just two ways in which the tracing of reasons can go on forever—either there is an infinite regress of reasons, or eventually the reasons go around in a circle. Thus a linear positive coherence theory must acknowledge that justified belief can result from either an infinite regress of reasons, or from circular reasoning.[61]

Both of these possibilities seem a bit puzzling, and foreign to the

---

60. For example, $\{P,(P \rightarrow Q)\}$ is a reason for Q.

61. Keith Lehrer (1990) and (1994) constructs a linear positive coherence theory in which justification can go around in a circle. The crucial belief making this possible is that one is a trustworthy cognizer. This belief is used in justifying the cognizer's other beliefs, and when asked for justification for this belief, Lehrer replies that he believes it and it is a trustworthy cognizer, so it is reasonable to believe that it is true. Laurence Bonjour (1985) has constructed a theory that he regards as a version of coherentism, and it looks initially like a linear coherence theory (see 117ff). But closer inspection reveals that he takes beliefs to the effect that one has a belief to be prima facie justified (this is what he calls "the doxastic presumption"), and hence in the taxonomy of this book his theory is a foundations theory.

classical view of reasoning that appears to underlie linear theories. Hence it is tempting to reject linear positive theories and instead adopt a holistic view of reasons. According to a *holistic* positive coherence theory, in order for S to have reason for believing P, there must be a relationship between P and the set of *all* of his beliefs (where this relationship cannot be decomposed into simple reason relationships holding between individual beliefs).[62] Note that on a holistic theory it is more natural to talk about "having reason" for holding a belief rather than "having *a* reason". In particular, it makes no sense to inquire about the reasons for one's reason for a belief. On this theory, one does not have *a* reason in the sense of a particular belief—rather, one has reason for a belief by virtue of his belief being appropriately related to his entire doxastic system.

The holistic/linear distinction can also be applied to negative coherence theories. A linear negative coherence theory is one taking all beliefs to be prima facie justified and adopting a "classical" view of the reasons and reasoning that can lead to the defeat of such prima facie justified beliefs. Alternatively, a holistic negative coherence theory will also take all beliefs to be prima facie justified but will insist that what defeats a person's justification for a belief is the relationship between that belief and the set of all the person's beliefs (where, again, this relationship cannot be decomposed into relationships between individual beliefs).

It is worth emphasizing that even linear theories can be more holistic than one might initially suppose. Linear theories are linear in the sense that the reason for one proposition is another proposition or a small set of propositions, but if reasons are defeasible then in an important sense what justifies a belief is not the individual proposition that is the reason but rather that proposition together with the absence of defeaters, and the latter appeals to all one's beliefs.

Holistic theories adopt a novel view of reasons according to which one's having reason for a belief is determined by the relationship of that belief to the whole amorphous structure of beliefs comprising his total doxastic system. Such a view of reasons contrasts with the more traditional view of reasons adopted by foundations theories and linear coherence theories, and although the vague picture of holistic reasons has intuitive appeal, it is a bit difficult to imagine just what they might be like. A concrete example of such a theory will be discussed in section three.

2.3 The Regress Argument

There are two familiar arguments that have been deployed repeatedly against coherence theories as a group. The first of these is the *regress argument*, which objects that coherence theories lead to an infinite regress of reasons and such a regress cannot provide justification.[63] This objection

---

62. Examples of holistic positive coherence theories are provided by Laurence Bonjour (1976), Gilbert Harman (1970), and Keith Lehrer (1974).
63. For discussion see Ernest Sosa (1980).

has often been regarded as fatal to all coherence theories, but in fact it is telling against only a few. It is simply false that all coherence theories lead to an infinite regress of reasons. This argument has no apparent strength against negative coherence theories, because they do not require reasons for beliefs. The regress argument really only bears upon some of the least plausible linear positive coherence theories. A linear positive coherence theory that identifies having P as a reason for believing Q with *having explicitly inferred Q from P* would run afoul of the regress argument, because it would require one to have performed infinitely many explicit inferences before one could be justified in believing anything, and that is presumably impossible. However, weaker understandings of what it is to have P as a reason for believing Q make the regress argument problematic, and holistic positive coherence theories would seem to avoid the regress argument altogether.

Although the regress argument fails to dispose of more than a few varieties of coherence theories, it is related to a more concrete objection that can be leveled against all linear positive coherence theories and seems to us to be fatal to them. Linear positive coherence theories look much like foundations theories as long as we focus our attention on beliefs not directly based upon perception. But the two types of theories diverge radically when we consider perceptual beliefs. According to foundations theories justification terminates at that point on epistemologically basic beliefs, but according to a linear positive coherence theory justification must instead loop back up to other "high level" beliefs. Foundations theories fail at this juncture because of their inability to handle perception, and linear positive coherence theories seem to fail for the same reason. The failure of foundations theories results from their attempt to base perceptual belief on epistemologically basic appearance beliefs, because we do not usually have such beliefs. Linear positive coherence theories seek to avoid this failure by taking perceptual beliefs to be based upon more ordinary physical-object beliefs, whose justification in turn rests upon other ordinary beliefs, and so on. This would indeed avoid the problem of not having appropriate epistemologically basic beliefs, but the difficulty with the proposed solution is that perceptual beliefs do not seem to be based, in the way envisioned, on physical-object beliefs either. For instance, suppose I see a book on my desk and judge perceptually that it is red. A typical foundations theory would allege that my reason for thinking that the book is red is my epistemologically basic belief that it looks red to me. But I normally have no such belief. A linear positive coherence theory proposes instead that my reason for thinking the book is red is some more ordinary physical-object belief. The trouble with this proposal is that there are no plausible candidates for such a reason. What could such a reason be? One suggestion we have heard is that our reason is the second-order belief that we believe the book to be red, but the claim that we ordinarily have such second-order beliefs is no more plausible than the foundationalist claim that we ordinarily have appearance beliefs. Furthermore, what could our reason be for the second order belief? Certainly not that we believe that we believe the book is red. Nor do there seem to be any other candidates with greater plausibility. The general difficulty is that perception is not inference from beliefs. When I believe on the basis of perception that the book is red, I do not infer that belief from something else that I believe. Perception is a causal process that inputs beliefs into our doxastic system without their being inferred from or justified on the basis of other beliefs we already have. This seems undeniable, and it appears to constitute a conclusive objection to all linear positive coherence theories.

2.4 The Isolation Argument

There is a second familiar argument that has often been leveled against coherence theories as a group. This is the *isolation argument*, which objects that coherence theories cut justification off from the world. According to coherence theories, justification is ultimately a matter of relations between the propositions one believes, and has nothing to do with the way the world is. But our objective in seeking knowledge is to find out the way the world is. Thus coherence theories are inadequate.[64]

There is a good point concealed in the isolation argument, but as stated, the argument is not compelling. The difficulty for the argument emerges when we ask for a clearer statement of the way in which coherence theories are supposed to cut justification off from the world. As with any doxastic theory, they make justification a function of what one believes, but what one believes is causally influenced by the way the world is. That is precisely what perception is all about. Perception is a causal process by virtue of which physical states of the world influence our beliefs. It might be objected that a merely causal relationship between our beliefs and the world is inadequate—what is required is some sort of *rational* relationship. But it is not at all clear how that would be possible. Rationality only seems to pertain to relations between beliefs. (This is the doxastic assumption.) Notice, in particular, that foundations theories do not differ from coherence theories with respect to the relationship between our beliefs and the world. The only way the world can influence our beliefs on either a foundations theory or a coherence theory is causally. Foundations theories and coherence theories differ only with regard to which beliefs are the causal progeny of our environment. Foundations theories suppose they are epistemologically basic appearance beliefs, while coherence theories suppose they are more ordinary physical-object beliefs having no privileged epistemic status. As far as we can see, the isolation argument provides no basis for preferring foundations theories to coherence theories.

---

64. We believe that the isolation argument was first formulated in Pollock (1974), although precursors to it can be found in the writings of C. I. Lewis. As should be apparent from the text, Pollock has subsequently abandoned it. Discussions of the isolation argument can be found in Laurence Bonjour (1976) and Susan Haack (1993).

Although, as stated, the isolation argument does not succeed in refuting coherence theories, there is a good point lurking beneath it. The point concerns doxastic theories in general, not just coherence theories, and it consists of a general difficulty in accommodating perception within any such theory. On a doxastic theory, the connection between our beliefs and the world can only be a causal one. But as difficult as this is to make clear, it seems that there should also be a rational connection of some sort. The reason for saying this is that we do distinguish between reasonable and unreasonable (justified and unjustified) perceptual beliefs. For instance, if a person sees a book, it looks red to him, and he judges that it is red, we will normally regard his belief that the book is red to be reasonable. But if he also knows that he is in a room bathed in red light and he knows the effect red light can have on apparent colors, then even if he is caused to believe that the book is red, we will not regard that belief as reasonable. Under the circumstances, given what he knows, he *should not* have come to believe that.

Doxastic theories can attempt to accommodate this intuition in various ways. In the end, we think that they are all unsuccessful. We will return to this objection at the end of the chapter where we will make it more precise and use it to motivate the adoption of a nondoxastic theory.

3. Holistic Positive
Coherence Theories

We have argued that linear positive coherence theories all fail because of their inability to produce plausible candidates for reasons for beliefs that result directly from perception. This failure turns specifically on the assumption that reasoning proceeds in terms of relations between individual beliefs. If we turn to holistic positive coherence theories, this objection no longer has any force. If the justifiedness of a belief is determined by its relationship to *all* other beliefs in a way that cannot be decomposed into linear reasons, then it is no longer clear whether we can be said to have a reason for perceptual beliefs. Holistic positive coherence theories require further exploration. There is only one such theory that has been carefully worked out, and that is due to Keith Lehrer (1974). Thus we will begin by giving a brief sketch of Lehrer's theory in order to give the reader some feeling for what holistic positive coherence theories are like. Then we will go on to make some more general remarks about holistic positive coherence theories as a group.

3.1 Lehrer's Theory

As a first approximation, Lehrer's proposal is that a belief P is justified for a person S if and only if for each belief with which P "competes", S believes that P is more probable than that competitor. To make the theory precise, we must make this notion of competition precise. We obviously want a proposition to compete with any proposition with which it is incompatible, but Lehrer argues that we must also allow propositions to compete with propositions with which they are logically consistent. He bases this on consideration of a lottery. Suppose S holds one ticket in a fair lottery consisting of 1000 tickets, each ticket having the same probability of being drawn, and S has true beliefs about all the relevant probabilities. The probability of any given ticket being drawn is thus one in a thousand. On this basis S may be tempted to conclude that his ticket will not be drawn. But note that S has equally good reason to draw the same conclusion about each ticket in the lottery. If he draws all those conclusions, the result is incompatible with something he already knows—namely, that some ticket will be drawn. Lehrer assumes that it cannot be reasonable to hold such an explicitly contradictory set of beliefs, but there is nothing to favor one of the tickets over any of the others, so he concludes that it cannot be reasonable to infer of *any* ticket that it will not be drawn. Consequently, Lehrer seeks to tailor his definition of competition so that his account of justification yields the result that in this example S is not justified in believing, of any ticket, that it will not be drawn.

Before examining Lehrer's definition of competition, we should take note of the assumption on which his argument is based. This is the assumption that one cannot be justified in holding an explicitly contradictory set of beliefs. Perhaps most epistemologists will endorse some principle such as this, but a few have explicitly rejected this principle.[65] We propose to simply grant Lehrer this assumption without further discussion so that we can give his theory a fair hearing.[66]

In order for his criterion of justification to yield the result that S is not justified in believing that any particular ticket will not be drawn, Lehrer wants to define competition in such a way that the proposition that a given ticket will lose competes with the proposition that any other ticket will lose. If these propositions are all in competition with each other, then because they are believed by S to be equally probable, S does not believe any of them to be more probable than each of its competitors. It follows by Lehrer's criterion that S is not justified in believing any of these propositions. This suggested to Lehrer that we should say that P competes with Q if and only if S believes Q to be "negatively relevant" to P. To say that Q is negatively relevant to P is to say that the probability of P *on the assumption that Q is true* is less than the probability of P without that assumption, that is, $prob(P/Q) < prob(P)$. This definition of competition will handle the lottery example, because any given ticket's not being drawn will slightly raise the probability of any other ticket being drawn and hence is negatively relevant to that other ticket's not being drawn.

---

65. See Richard Foley (1992a) and (1992b).

66. We do not want to give the impression that Lehrer is unaware of the contentiousness of the consistency assumption. He explicitly defends it in his (1975).

Lehrer further illustrates his analysis by considering the perceptual judgment that he sees a red apple. He observes that he not only believes that he sees an apple, but he also believes that the proposition that he sees a red apple is more probable than the proposition that he sees a wax imitation, or a painting of an apple, or that he is hallucinating, and so forth. All of the latter are negatively relevant to his actually seeing a red apple and hence are in competition with that proposition, but in each case he believes them to be less probable.

This is only a crude sketch of Lehrer's theory. He goes on to argue that a number of complicated emendations are required in order to make the theory work, but we will not pursue them here. This at least gives the flavor of the theory. It is an ingenious theory, and it is the only attempt to actually construct a precise holistic positive coherence theory that can be found in the contemporary literature. Lehrer's theory illustrates how the justifiedness of a belief might be a function of one's total doxastic system without being determined by relations between individual beliefs of the sort envisioned by "classical" theories of reasons and reasoning.[67]

3.2 Problems for Holistic Positive Coherence Theories

Holistic positive coherence theories constitute an interesting, although largely unexplored, category of epistemological theory. But despite their intuitive appeal, there are some general difficulties that infect them as a group. One kind of problem is analogous to a problem we encountered in trying to construct a plausible foundations theory. The picture underlying coherence theories is of a vast web of interrelated beliefs, but this picture becomes dubious when we reflect upon the distinction between occurrent and non-occurrent beliefs. The picture may give an accurate portrayal of our total doxastic system if we construe it as consisting of occurrent and non-occurrent beliefs alike, but it is quite inaccurate as a portrayal of our occurrent beliefs alone. What must be asked is whether non-occurrent beliefs can play a direct role in justification. It certainly *seems* that beliefs should be relevant to justification only insofar as one recalls them. Non-occurrent beliefs can be recalled with varying degrees of difficulty, and if one is at the moment unable to recall a particular non-occurrent belief then it does not seem that that belief should play any role in determining whether some other occurrent belief is justified.

In deciding whether this objection has any real strength, we must give some thought to the concept of epistemic justification. Epistemic justification is supposed to be "regulatory" in some sense. It is concerned with which beliefs we should hold and which we should not. Considerations of epistemic justification are supposed to guide us, somehow, in deciding what to believe. It seems that a consideration can guide us in this way only if we have access to it. This is the same intuition that

---

67. For more recent incarnations of Lehrer's views, see his (1981) and (1982). He has defended different kinds of coherence theories in more recent publications (1990) and (1994).

underlies the doxastic assumption. It seems that we do not have access, in the requisite sense, to non-occurrent beliefs, and so they cannot be relevant to justification.

It is hard to assess this objection. For one thing, it seems to turn upon an illegitimate kind of "doxastic voluntarism"—we do not literally *decide* what to believe. We do not have voluntary control over our beliefs. We cannot just decide to believe that 2+2 = 5 and thereby do it. We have at most indirect control over what we believe. We can try to get ourselves to believe something by repeatedly rehearsing the evidence for it, or putting countervailing evidence out of our minds, or by deliberately seeking new evidence for it, but we cannot voluntarily make ourselves believe something in the same sense that we can voluntarily clench our fists. This makes it hard to understand how epistemic norms and considerations of epistemic justification can play a regulatory role in belief. If we cannot directly control what we believe, how can we regulate it in accordance with epistemic norms? On the other hand, epistemic justification is definitely a normative notion, and we do raise questions about whether we *should* hold various beliefs. Some kind of regulation seems to be involved, but it is hard to see what it amounts to.[68]

The way in which epistemic norms regulate belief will become clearer in chapter five. In the meantime it is hard to evaluate the objection that, because we only have access to those of our beliefs that are occurrent, non-occurrent beliefs can play no direct role in epistemic justification. There is something intuitively persuasive about this; if it is correct then it seems to create difficulties for holistic coherence theories (positive or negative) because there will not be enough material available for a coherence relation to appeal to. Taking Lehrer's theory as an example, suppose S believes P, which competes with $Q_1,...,Q_n$. S may non-occurrently believe that P is more probable than each Q_i (even this is dubious—do people normally have such beliefs?), but it is quite unlikely that in a normal situation S will have any occurrent probability beliefs of this sort. Thus if we are only allowed to appeal to occurrent beliefs, it will result from Lehrer's theory that we are hardly ever justified in believing anything.

This objection has a certain amount of intuitive force, but it is hard to be sure whether it should be regarded as fatal to Lehrer's theory. What this underscores is a problem that will become central when we turn to nondoxastic theories. This is the problem of understanding what epistemic justification is all about. Thus far we have taken the notion as given, we have assumed that we have a rough understanding of it, and we have formulated epistemological theories in terms of it. But what we are finding is that our grasp of the concept of epistemic justification is unclear in important respects, and that makes it difficult to adjudicate disputes between rival epistemological theories. Furthermore, it is not satisfactory to leave the central concept of epistemology unanalyzed. Only by pro-

---

68. For more on doxastic voluntarism, see William Alston (1988).

viding some sort of analysis or clarification of the concept can we ultimately resolve these disputes. Unfortunately, providing such an analysis has proven to be a very hard problem. Most epistemologists have remained mute on the subject, not because they have no interest in it but because they have no answers to propose. The only epistemologists to propose analyses have been the externalists, and we will examine their accounts in chapter four. We will proffer our own analysis in chapter five.

The preceding objection is tempting but inconclusive. We can raise a different sort of objection to holistic positive coherence theories that seems more conclusive. This concerns the problem of the basing relation, which was discussed briefly in chapter one. There is a distinction between having good reason for believing something (perhaps without appreciating the reason) and believing something *for* a good reason. This is, roughly, the distinction between a justi*fied* belief and a justi*fiable* belief. A justifiable belief is one the believer could become justified in believing if he just put together in the right way what he already believes. To illustrate, a woman might have adequate evidence for believing that her husband is unfaithful to her, but systematically ignore that evidence. However, when her father, whom she knows to be totally unreliable in such matters and biased against her husband, tells her that her husband is unfaithful to her, she believes it on that basis. Then her belief that her husband is unfaithful is unjustified but justifiable.

Any correct epistemological theory must allow this distinction. The problem with holistic positive coherence theories is that they seem to be incompatible with the distinction. The basing relation is at least partly a causal relation.[69] Being justified in holding a belief on a certain basis consists of your belief "arising out of" that basis in some appropriate way. But this does not seem to be possible on a holistic positive coherence theory. In order for the notion of a justified (as opposed to a merely justifiable) belief to make sense within such a theory, the coherence relation (whatever it is) must be such that the belief's cohering with one's overall doxastic system *can* cause one (in an appropriate way) to hold the belief. The coherence relation must be "appropriately causally efficacious" in the formation of belief. There are just two possibilities for the nature of the causal chains leading from coherence to belief. On the one hand, they might be "doxastic", whereby the believer first comes to believe that P coheres with his other beliefs and then comes to believe P on that basis. The simple objection to such a doxastic reconstruction of the basing relation is that we do not ordinarily have any such beliefs about coherence. A second objection is that it seems to lead to an infinite regress—would we not have to be justified in believing that P coheres,

---

69. Keith Lehrer (1971) and (1990) has argued against this, but we do not find his counterexample persuasive. There have been many attempts in the literature to show the intuitiveness of understanding the basing relation as partly causal. See Gilbert Harman (1973), William Alston (1976), Alvin Goldman (1979), (1985), and (1986), Marshall Swain (1981), and Hilary Kornblith (1980).

and if so would that not require our first coming to believe that the belief that *P* coheres with our other beliefs coheres with our other beliefs? We think it is clear that a holistic positive coherence theory cannot adopt such a doxastic account of the basing relation. Consequently, it must be possible for the kind of causal connection involved in the basing relation to cause belief in *P* without the believer having any beliefs about whether *P* coheres with his other beliefs. But is it at all plausible to suppose that the coherence relation is such that P's cohering with the believer's other beliefs can cause belief in a nondoxastic way? This must depend upon the nature of the proposed coherence relation, but it is hard to see how any plausible coherence relation could be appropriately causally efficacious.

Consider, for example, Lehrer's proposed coherence relation. If coherence consists of S's believing *P* to be more probable than each of its competitors (individually, not collectively), how *could* this cause *S* to believe *P* except via S's first coming to believe that *P* coheres in this way? The same thing would seem to be true for any plausible coherence relation. Without concrete candidates to examine it is hard to be absolutely sure, but it seems that coherence relations will always involve elaborate logical relationships between beliefs, and the holding of such relationships can only be causally efficacious by virtue of one's coming to *believe* that the relationships hold. As a matter of psychological fact, such elaborate relationships cannot be nondoxastically causally efficacious. It seems quite unlikely that there could be any holistic coherence relation that can be appropriately causally efficacious in belief formation in such a way as to allow us to distinguish between justified and merely justifiable belief. This argument is not conclusive, because we cannot examine all possible putative coherence relations, but it provides a strong reason for being suspicious of holistic positive coherence theories as a group, and it seems to provide a conclusive reason for rejecting concrete examples of such theories.

The preceding arguments at least *suggest* the rejection of all positive coherence theories. If any coherence theory is to be defensible, it must be a negative coherence theory, so we turn to those theories next.

4. Negative Coherence Theories

Negative coherence theories accord all beliefs the status of prima facie justification. A negative coherence theory tells us that we are automatically justified in holding any belief we do hold unless we have some positive reason for thinking we should not hold it.[70] Something like the

---

70. This is what Gilbert Harman (1984) and (1986) calls "the principle of positive undermining".

Neurath metaphor provides the principal motivation for such a theory. The idea is that we must start with the stock of beliefs we already have and then amend those beliefs in light of themselves.

Negative coherence theories differ in their account of what can defeat the justification of a belief. Perhaps the most natural account is one adopting the "classical" picture of reasons embodied in the foundations theory described in chapter two. However, Gilbert Harman (1973, 1980, 1984, and 1986) is the only author to argue vigorously for a negative coherence theory, and he adopts a nonclassical view of defeaters.

4.1 Harman's Theory
Harman (1984) writes:

> Reasoning is a process of change in overall view. One's conclusion is in a sense one's whole modified view and not any single statement. ... Reasoning involves two factors—conservatism and coherence. One seeks in reasoning to minimize change in one's overall view, modifying it only to explain more and leave less unexplained. (154)

According to Harman, coherence is *explanatory coherence*. We modify our overall doxastic system only in response to explanatory considerations. We may add a belief because it explains other things we believe, and we may delete a belief because we cannot explain how it could be true or because it is incompatible with something else we believe on explanatory grounds.

Harman's theory illustrates something interesting about negative coherence theories. By definition, negative coherence theories hold that reasons are not required for the justification of a belief—beliefs are automatically justified unless you have a reason for rejecting them. This gives reasons a negative role in belief change. On the surface, this is compatible with their also playing a positive role. It may be that reasons also function to justify the acquisition of new beliefs on the basis of beliefs already held. Harman's theory seems to incorporate both sorts of elements, telling us that we should add beliefs under some circumstances and delete beliefs under others. According to Harman, we need reasons to accept beliefs but we do not need reasons to retain them. Harman's theory is subject to criticism on this ground. A negative coherence theory can never assign reasons more than a negative role. It might seem that they could give reasoning the positive role of making it permissible to acquire new beliefs under various circumstances. After all, this would be the positive role accorded to reasoning by either a foundations theory or a positive coherence theory. But a negative coherence theory cannot adopt this line. That is because according to a negative coherence theory it is *always* permissible to adopt a new belief—*any* new belief. Because beliefs are prima facie justified you do not need a reason for adopting a new belief. Of course, having adopted a new belief it may immediately become obligatory to reject it (because it is incoherent with your other beliefs). But having positive reasons for the new belief could only ensure that it is not obligatory to reject it if the positive reasons are constructed so as to automatically ensure that there are no negative reasons legislating the rejection. In that case (assuming a negative coherence theory), the positive reason is really playing no role other than the negative one of ruling out reasons for rejecting the new belief. In other words, it is still just a defeater. So within a negative coherence theory, a positive role cannot be assigned to reasons.

This suggests modifying Harman's theory a bit by assigning only negative roles to explanatory coherence, taking a belief to be unjustified if either (1) we cannot (within the context of our overall doxastic system) explain how it could be true, or (2) the best explanation for other features of our beliefs is incompatible with this particular belief. Harman defends these principles of belief revision by arguing that as a matter of psychological fact people do not stop believing things whenever they cease to have reasons for believing them. On the contrary, a principle of conservativeness operates in belief revision, leading us to reject beliefs only when we acquire positive reasons for doing so. His argument for this is, roughly, that we do not usually remember our reasons for very long, and when we forget them we persist in holding the belief anyway. Furthermore, that is the only reasonable way for a creature with limited memory storage to function. We do not have the capacity to remember all our reasons. Normally, what is important is the conclusions—not the reasons. All of this seems right, but as we pointed out in the introduction to this chapter, such considerations do not constitute a defense of negative coherence theories. This is because they can be accommodated just as well by according memory a positive role in epistemic justification, supposing that one's seeming to remember something constitutes a positive reason for believing it. Harman is certainly right that a principle of conservativeness operates in our system of belief revision, but this can be accommodated *either* by adopting a negative coherence theory *or* by adopting a positive view of reasons and taking apparent memory to supply positive reasons. To decide between these two accounts we must examine them more closely.

4.2 Problems with Negative Coherence Theories

One of us (Pollock) always found negative coherence theories appealing, and has repeatedly returned to the task of trying to construct a defensible theory of this sort. But despite their intuitive appeal, we are convinced that such theories cannot be successfully defended. There are two related sorts of considerations that lead us to this conclusion. The first is analogous to one of the objections we raised to holistic positive coherence theories. Any reasonable theory must make it possible to draw the distinction between justified and merely justifiable beliefs. But that distinction seems to be incompatible with a negative coherence theory. The distinction has to do with the basing relation and the basis upon which one holds a belief. But if reasons do not play some positive role in belief formation, there can be no such thing as holding a belief for a particular reason or on a particular basis. If reasons play only a negative role, then they are only relevant as defeaters for beliefs already held. According to a negative coherence theory, how one comes to hold a belief is irrelevant to its justification. But this seems clearly false. The examples used earlier to illustrate the distinction appear to demonstrate conclusively that reasons do play a positive role in the justification of belief. For instance, recall the example in chapter two of the person who is constructing a mathematical proof, who wants to draw a particular conclusion at a certain point, who has earlier lines in her proof from which that conclusion follows immediately, but who overlooks that connection and in despair simply writes down the desired conclusion thinking to herself, "Oh, that's got to be true." That person is not justified in believing the conclusion, and the reason she is not justified is that she does not believe it on the basis of the earlier lines from which it follows. Those earlier lines would give her a good reason if she just saw the connection, but she does not and so she is unjustified in believing the new conclusion.

When one believes something by remembering it, positive and negative accounts of reasoning give the same result. The above objection amounts to a defense of the principle that if one holds a non-memory belief for a bad reason or for no reason at all, that belief is not justified. This is precisely what a positive account of reasons would predict. As we remarked in chapter two, there seems to be just one way that a negative account can accommodate this, and that is by supposing that when one holds a belief for a bad reason or for no reason at all, that automatically gives him a reason for thinking he should not hold the belief. In point of fact, the negative coherence theorist cannot formulate the supposition in this way because he does not accord reasons a positive role, but he might instead formulate the supposition as follows:

If one believes P unjustifiably then he automatically has a belief that is a defeater for believing P.

But does one automatically have such a belief? Thus far we have avoided simply rejecting this epistemic principle, but our only reason for doing so was to make doxastic theories more plausible for the sake of the discussion. In fact, this epistemic principle seems obviously wrong. This is because one need not be aware of why one holds a belief. Upon reflecting on a belief I may decide that I have no good reason for holding it and that it is just a manifestation of wishful thinking or some other epistemically proscribed cognitive process. Once I have decided that, it seems I have a reason for thinking I should not hold the belief, but prior to deciding that it does not seem that I have any such reason. Nevertheless, if I hold the belief on the basis of wishful thinking, then my belief is unjustified. But then it follows that the belief can be unjustified without my having a defeating belief (or more intuitively, without my having a reason for thinking I should not hold the belief). In other words, not all beliefs are prima facie justified, and hence negative coherence theories are false.

These two objections seem to be conclusive. Reasons must be accorded some sort of positive role in justification. The intuition to the contrary turns upon confusing memory belief with believing something for no reason. These are two quite different phenomena. In light of the preceding discussion, it seems inescapable that memory itself plays a certifying role in memory beliefs. The belief is justified *because* we seem to remember it. On the other hand, the certifying role of memory is no more easily accommodated by foundations theories than it is by coherence theories. This is because, as we noted in chapter two, when we remember something we do not usually have a belief to the effect that we seem to remember it, and hence the memory belief cannot be regarded as being held on the basis of an epistemologically basic belief about seeming to remember. This difficulty is a very general one. We regard it as symptomatic of the failure of all doxastic theories. Doxastic theories are based on the doxastic assumption according to which the only things relevant to the justifiedness of a belief are one's beliefs. Neither perception nor memory can be accommodated within a doxastic theory. Arguing that will be the burden of the next section.

5. Nondoxastic Theories and Direct Realism

Perception is a causal process that introduces beliefs about physical objects into our doxastic system. Given the doxastic assumption, the only belief changes that are subject to rational epistemic evaluation are those made exclusively in response to your other beliefs. Facts about the perceptual situation, such as how you are appeared to, can only be relevant to the epistemic evaluation of your perceptual belief insofar as you belief those facts. In a normal case you have no beliefs about how you are appeared to, and the beliefs you do have prior to acquiring the perceptual belief are not sufficient to uniquely determine what perceptual belief you should acquire. For example, the beliefs you have antecedently will not normally determine whether, upon examining a new object, you should believe it to be red or green. You normally have to look at an object to determine what color you should believe it to be. In other words, perception is not inference from antecedently held beliefs. Because perceptual beliefs cannot inferred from other beliefs, there is no way to "consider a potential perceptual belief" before we acquire it and make a rational decision whether to adopt it. We can, of course, consider coun-terfactually whether a certain belief would cohere if we were to adopt it, but that cannot be what transpires in perception because we do not know which potential perceptual belief to evaluate until we actually have the belief. "There is a red book before me" and "There is a green book before me" may both cohere with the rest of my beliefs. It is not coherence that determines which to evaluate—it is the causal processes of perception that inject one of these beliefs rather than the other into my doxastic system. It would only be possible to evaluate a potential perceptual belief "before the fact" if we had appearance beliefs and were deciding whether to adopt the perceptual belief in response to them. In other words, it is only possible to evaluate potential perceptual beliefs about physical objects prior to acquiring them if we can reduce perception to inference from epistemologically basic beliefs, and we have seen that that cannot be done.

Although it follows from the doxastic assumption that the perceptual acquisition of beliefs about physical objects cannot be subject to epistemic evaluation, the beliefs themselves can be subject to epistemic evaluation in terms of how they are related to other beliefs. But this can only happen *after* they become beliefs. Thus if we accept the doxastic assumption, we must view the perceptual acquisition of beliefs as automatically epistemically permissible and hence the beliefs themselves are automatically justified *unless* they conflict somehow with other beliefs. That is, they must be prima facie justified.

Apparently, any doxastic theory must regard the beliefs issuing most directly from perception as prima facie justified. (The same point can be made about memory beliefs.) Doxastic theories can differ with regard to which beliefs are the ones issuing most directly from perception, and how generally the property of prima facie justification is shared by other beliefs, but they must accord this status to whichever beliefs they identify as perceptual beliefs.[71] Stereotypical foundations theories assume that the beliefs issuing most directly from perception are appearance beliefs (and they typically assign them an even stronger status than prima facie justification), but that seems to be psychologically inaccurate. The normal doxastic progeny of perception are perfectly ordinary beliefs about physical objects. A psychologically realistic foundations theory must maintain that physical-object beliefs are prima facie justified while most other beliefs are not. We argued at the end of chapter two, however, that such a position is untenable. The difficulty was that one can believe the very same thing either perceptually or for a variety of non-perceptual reasons, and if one holds the belief for a non-perceptual reason and it is a bad reason then one is not justified in holding the belief. It follows that the

---

71. The way Laurence Bonjour (1985) proposes to accomodate perceptual beliefs is to claim that they are "cognitively spontaneous" and to stipulate that we must show a special deference to these beliefs in our inferences (he calls this the "observation requirement"). The way to show special deference is by reasoning from the premises that I now have a cognitively spontaneous belief and that cognitively spontaneous beliefs of a certain kind are likely to be true, to the conclusion that the contents of the cognitively spontaneous belief are likely to be true. This proposal runs directly into the difficulties we note here. It also requires apparently foundational beliefs to the effect that one has various cognitively spontaneous beliefs.

doxastic assumption must be false.

This argument against the doxastic assumption is of vital importance, so let us briefly repeat it stripped of explanatory remarks:

1. Suppose the doxastic assumption holds. Then the justifiedness of a perceptual belief can only depend upon your other beliefs and cannot depend upon any features of perception not encoded in belief. In particular, it cannot depend upon your being appeared to in appropriate ways if (as is usually the case) you do not have beliefs to the effect that you are appeared to in those ways.

2. In a normal case, your other beliefs cannot determine that just one possible perceptual belief could be justified. You cannot, for example, tell what color you should believe something to be without looking at it. Therefore, perceptual beliefs cannot be evaluated before being acquired, because it is the very fact of acquiring the perceptual belief that determines which possible belief (e.g., "That is red" rather than "That is green") to evaluate.

3. It follows that the acquisition of a perceptual belief is automatically justified, and its relationship to other beliefs can only be relevant negatively. In other words, perceptual beliefs must be prima facie justified.

4. Perceptual beliefs are ordinary physical-object beliefs, and such beliefs can also be held for non-perceptual reasons. If such reasons are bad reasons, the beliefs are not justified. But then it follows that they are not prima facie justified.

5. (4) conflicts with (3), so the assumption from which (3) followed, namely the doxastic assumption, must be false.

The falsity of the doxastic assumption can also be argued more straightforwardly by looking at physical-object beliefs held directly on the basis of perception. The doxastic assumption implies that the acquisition of such beliefs is not subject to epistemic evaluation—the only thing we can evaluate epistemically is one's continuing to hold such a belief after it is acquired. On the doxastic assumption we must adopt the belief first and then decide whether to discard it rather than deciding whether to adopt it in the first place. But this is manifestly false. For example, suppose you know you are in a room bathed in red light (e.g., this is done in military maneuvers to enhance night vision), and you know the effect this has on the colors things look. Under these circumstances, if you see a piece of paper before you and it looks red, you would be unjustified in making the perceptual judgment that it is red. It is not just that you would be unjustified in retaining that belief once you acquire it—you should not acquire it in the first place. You "know better". This is a normative epistemic judgment. The possibility of such a judgment indicates that epistemic norms apply to the perceptual acquisition of beliefs, and hence must be able to appeal to nondoxastic states. What makes a normal perceptual belief epistemically permissible (i.e., justified) is that one *is in* an appropriate perceptual state (i.e., is appeared to in appropriate ways) and has no defeating beliefs. The latter is a matter of beliefs, but the former is not. One does not have to *believe* that he is in the perceptual state. So the justifiedness of a belief is a function of more than just one's beliefs. At the very least, how one is appeared to is also relevant. The same point can be made about memory. The fact that one seems to remember can make one justified in holding a belief—one does not have to believe that he seems to remember. It follows that the doxastic assumption is false.

To recapitulate, foundations theories and coherence theories fail for basically the same reason—epistemic rationality is not just a function of one's beliefs. Both beliefs and nondoxastic perceptual and memory states are relevant to the justifiedness of a belief. Foundations theories try to take account of perceptual states and memory states by supposing we always have beliefs about them, but in this they are mistaken. Coherence theories rightly reject such epistemologically basic beliefs, but they throw out the baby with the bathwater. By rejecting appearance and apparent-memory beliefs but refusing to allow nondoxastic appeal to perceptual states and memory states they make it impossible to accommodate perception and memory. The right response is to reject appearance beliefs, in the sense of admitting that the justifiedness of perceptual and memory beliefs does not depend upon our having appearance and apparent-memory beliefs, but to acknowledge that our epistemic norms must be able to appeal to perceptual and memory states directly and without doxastic mediation. It is *the fact that* I am appeared to redly that justifies me in thinking there is something red before me, and it is *the fact that* I seem to remember that Columbus landed in America in 1492 that justifies me in believing that. It is not my *believing* that I am appeared to redly or my *believing* that I seem to remember that Columbus landed in 1492 that is required for justification.

The true epistemological theory must be nondoxastic. This still leaves us with a wide variety of possibilities. In particular, there are both internalist and externalist nondoxastic theories. Internalist nondoxastic theories insist that although the justifiedness of a belief is not a function exclusively of one's other beliefs, it is a function exclusively of one's internal states, where the latter include both beliefs and nondoxastic states. Internal states are, roughly, those to which we have "direct access". Externalist theories insist, on the other hand, that the justifiedness of a belief may also be determined by entirely external considerations like the reliability (in the actual world) of the cognitive processes producing the belief.

Externalist theories depart radically from traditional doxastic theories, and they will be the subject of the next chapter. Nondoxastic internalist theories, however, may have structures very similar to classical foundations theories. For instance, *direct realism* is the view that perceptual states can license perceptual judgments about physical objects directly and without mediation by beliefs about the perceptual states. Direct realism can have a structure very much like a foundations theory. Our own view is that the foundations theory sketched in chapter two gets things almost right. Where it goes wrong is in adopting the doxastic assumption and thereby assuming that perceptual input must be mediated by epistemologically basic beliefs. It now seems clear that epistemic norms can appeal directly to our being in perceptual states and need not appeal to our having beliefs to that effect. In other words, there can be "half-doxastic" epistemic connections between beliefs and nondoxastic states that are analogous to the "fully doxastic" connections between beliefs and beliefs that we call "reasons". We propose to call the half-doxastic connections "reasons" as well, but it must be acknowledged that this is stretching our ordinary use of the term 'reason'. The motivation for this terminology is that the logical structure of such connections is analogous to the logical structure of ordinary defeasible reasons. That is, the half-doxastic connections convey justification defeasibly, and the defeaters are completely analogous to the defeaters proposed by the foundations theory formulated in chapter two.

   Direct realism retains the attractive intuitions about the connection between justification and reasoning that are part and parcel of classical foundations theories, while avoiding the shortcomings of such theories by giving up the doxastic assumption. We are inclined to regard some form of direct realism as true. We have not yet argued for this conclusion, however. The basic argument so far has been against the doxastic assumption, and nondoxastic theories other than direct realism are possible. In particular, there are externalist nondoxastic theories. We will turn to the consideration of externalist theories in the next chapter. Finally, in chapter five, we will argue that a correct understanding of the nature of epistemic norms requires that a true epistemological theory be internalist. As it must also be nondoxastic, we will take that to be a defense of direct realism, and we will give a somewhat more detailed sketch of how direct realism should be formulated.

4
EXTERNALISM

1. Motivation

All of the theories discussed so far have been internalist theories. Doxastic theories take the justifiability of a belief to be a function exclusively of what else one believes. Internalist theories in general loosen that requirement, taking justifiability to be determined more generally by one's internal states. Beliefs are internal states, but so are perceptual states, memory states, and so on. Internal states have been only vaguely characterized as those to which we have "direct access". This vagueness must eventually be remedied, but further clarification will have to await later chapters. Externalist theories loosen the requirement for justifiability still further, insisting that more than the believer's internal states is relevant to the justifiability of a belief. For instance, reliabilism takes the reliability of the cognitive process generating the belief to be relevant to its justifiability.

The primary motivation for most externalist views proceeds in two stages. The first consists of the rejection of all doxastic theories. That rejection was defended above on the grounds that doxastic theories cannot handle perceptual input, which is basically a nondoxastic process that is nevertheless subject to rational evaluation. Thus we must select an epistemological theory from among nondoxastic theories. The externalist proposes that this selection be driven by a particular intuition. This is the intuition that we want our beliefs to be probable—we should not hold a belief unless it is probably true. Probability, however exactly it is brought to bear on the selection of beliefs, is an external consideration. The probability of one's belief, or the reliability of the cognitive process producing it, is not something to which one has direct access. Thus we are led to an externalist theory that evaluates the justification of a belief at least partly on the basis of external considerations of probability.

There is also a secondary motivation for the particular kinds of externalist theories that have been proposed in the literature. The doxastic theories discussed above propose very elaborate criteria for the justification of a belief. For instance, both the foundations theory of chapter two and the version of direct realism sketched at the end of the last chapter proceed by laying down a complex array of epistemic rules governing the justification of various kinds of judgments. A linear coherence theory proceeds similarly, adopting a structure of reasons similar to those involved in foundations theories (although in the case of negative coherence theories the rules only concern defeaters). And a similar point can be made about extant holistic coherence theories. For example, although Lehrer's theory does not proceed in terms of linear reasons, he too proposes a very complicated criterion for justification. (The full complexity of his criterion is not indicated by the brief sketch of his theory given in chapter three.) All of these theories proceed by initially taking the concept of epistemic justification for granted, and then using our intuitions about epistemic justification to guide us in the construction of a criterion that accords with those intuitions. A telling objection can be raised against all of these internalist theories—they are simultaneously ad hoc and incomplete. They are ad hoc in that they propose arrays of epistemic rules without giving any systematic account of why those should be the right rules, and they are incomplete in that they propose no illuminating analysis of epistemic justification.[72] The methodology of internalism has been to describe our reasoning rather than to justify it or explain it. These two points are connected. As long as we take the concept of epistemic justification to be primitive and unanalyzed, there is no way to *prove* that a particular epistemic rule is a correct rule. All we can do is collect rules that seem intuitively right, but we are left without any way of justifying or supporting our intuitions.

It might be responded that internalist theories do not leave the concept of epistemic justification unanalyzed. The criteria of justifiedness that they propose could be regarded as analyses of justification.[73] But even if one of these criteria correctly described which beliefs are justified, it would not explain what epistemic justification is all about. The criterion would not provide an *illuminating* analysis. Because of its very complexity we would be left wondering why we should employ such a concept of epistemic justification. Its use would be unmotivated. What we would have is basically an ad hoc theory that is contrived to give the right answer but is unable to explain in any deep way why that is the right answer. What we really want is an analysis of epistemic justification that makes it manifest why we should be interested in the notion. The analysis could then be used to generate a principled account of epistemic norms. An account of this sort would not proceed by simply listing the epistemic rules that seem to be required to license those beliefs we regard as actually being justified. Instead, it would *derive* the rules from the concept of epistemic justification. No epistemological theory yet discussed in this book has that character.

Most doxastic theorists have had remarkably little to say about the analysis of epistemic justification. This is not due to lack of interest—it is due to lack of ideas. Within the context of an internalist theory, the analysis of epistemic justification has proven to be an extremely difficult problem. Here is a respect in which externalist theories appear to have a marked advantage. If it is granted that the justifiedness of a belief can be determined partly by external considerations, then it becomes feasible to

---

72. Ernest Sosa (1980) raises this objection.
73. For example, Lehrer casts his theory as an analysis of epistemic justification.

try to analyze justification in terms of probability. That seems like a very hopeful approach. Lehrer (1974) expresses this idea succinctly in his critical discussion of foundationalism:

> What I object to is postulation without justification when it is perfectly clear that an unstated justification motivates the postulation. The justification is that people are so constructed that the beliefs in question, whether perceptual beliefs, memory beliefs, or whatnot, have a reasonable probability of being true. (77)

It is worth noting, however, that Lehrer's own theory diverges from this basic idea and thereby becomes subject to the same sort of criticism. This is because technical considerations lead him to an analysis no less complex than other internalist analyses. The simple intuition that justified beliefs must be probable provides no explanation at all for this complicated structure. Lehrer has merely replaced the complicated structure of foundations theories by another complicated structure, without any explanation for why justification should have either structure. He is as guilty of postulating epistemic rules as the foundations theorist.[74]

The hope that some simple analysis can be given of epistemic justification in terms of probability has had a powerful influence on epistemologists and has made externalist theories seem attractive. But it must be emphasized that such an analysis has to be simple. If the externalist is led to a complicated analysis, he will have been no more successful than the internalist in explaining why epistemic justification is a notion of interest to us. His theory will be no less ad hoc. The only virtue of the externalist theory will be the same one claimed (perhaps falsely) by all other theories, namely, that they correctly pick out the right beliefs as justified. Of course, that itself would be no mean feat.

To summarize, there are two sources for the appeal externalism has exercised on recent epistemologists. On the one hand, externalist theories seek to capture the common intuition that there is an intimate connection between epistemic justification and probability. On the other hand, there is the hope that an externalist analysis can explain what justification is all about rather than merely providing a correct criterion for justifiedness.

Externalist theories promise dividends not provided by any of the theories thus far discussed. But to evaluate these promissory notes, we must look more carefully at actual externalist theories. Two major kinds of externalist theories can be found in the literature—probabilism and reliabilism. Probabilism attempts to characterize the justifiedness of a belief in terms of its probability and the probability of related beliefs. Reliabilism seeks instead to characterize the justifiedness of a belief in terms of more general probabilities pertaining not just to the belief in

---

74. Lehrer's theory diverges from his basic intuition in another respect as well. His theory is formulated in terms of *beliefs about* probabilities rather than in terms of the probabilities themselves. Otherwise, his theory would not be a doxastic theory.

question but to the cognitive processes responsible for the belief. Among epistemologists, reliabilism is a more familiar version of externalism. Since reliabilism is partially predicated on probabilities, however, we will discuss probabilism first. Thus, probabilism and reliabilism will be the subjects of sections three and four, respectively. Before we can discuss them, however, we must lay some groundwork.

2. Varieties of Probability

Philosophers tend to make too facile a use of probability. They throw the word around with great abandon and often just assume that there is some way of making sense of their varied pronouncements, when frequently there is not. The main difficulty is that there is more than one kind of probability and philosophers tend to conflate them. A reasoned assessment of externalist theories requires us to make some careful distinctions between different kinds of probability. One important distinction is that between physical probability and epistemic probability. *Physical probability* pertains to the structure of the physical world and is independent of knowledge or opinion. For instance, the laws of quantum mechanics state physical probabilities. Physical probabilities are discovered by observing relative frequencies, and they are the subject matter of much of statistics. Physical probability provides the stereotype in terms of which most philosophers think of probability. But another important use of the word 'probable' in ordinary speech is to talk about degree of justification. For instance, after looking at the clues the detective may decide that it is probable that "the butler did it". Probability in this sense is directly concerned with knowledge and opinion and has no direct connection to the physical structure of the world. The *epistemic probability* of a proposition is a measure of its degree of justification. Epistemic probability is relative to a person and a time. It is an open question whether numerical values can be assigned to epistemic probabilities, and even if they can it is not a foregone conclusion that they will conform to the same mathematical principles (the probability calculus) as physical probabilities.

In addition to physical and epistemic probabilities, it is arguable that there are mixed physical/epistemic probabilities that are required for decision theory, weather forecasting, and so on. These probabilities appeal both to general physical facts about the world and to our knowledge of the present circumstances and how they relate to those general physical facts.[75] We will say more about these mixed probabilities below.

There is a second distinction that is related to but not identical with the distinction between physical and epistemic probabilities. We can

---

75. For more on the interrelationships between these various kinds of probability, see Pollock (1984a), and for a complete account see Pollock (1989).

distinguish between *definite probabilities* and *indefinite probabilities*. Definite probabilities are the probabilities that particular propositions are true or that particular states of affairs obtain. Indefinite probabilities, on the other hand, concern concepts or classes or properties rather than propositions. We can talk about the indefinite probability of a smoker contracting lung cancer. This is not about any particular smoker—it is about the class of all smokers, or about the property of being a smoker and its relationship to the property of contracting lung cancer. Some theories of probability take definite probabilities to be basic, and others begin with indefinite probabilities. Epistemic probabilities are always definite probabilities because they reflect a degree of belief in a particular proposition. Physical probabilities might be either. Theories taking physical probabilities to be closely related to relative frequencies make them indefinite probabilities. This is because relative frequencies concern classes or properties rather than single individuals. But there is also an important class of theories—"propensity theories"—that take the basic physical probabilities to pertain to individual objects. For example, we can talk about the probability that a particular coin will land heads on the next toss. Such "propensities" are definite probabilities.

With these preliminary distinctions before us, let us turn to some developed proposals regarding probability. There are three broad categories of probability theories to be found in the current literature, and externalists could in principle appeal to any of them, so we turn now to a brief sketch of each kind of theory.

2.1 Subjective Probability

Theories of subjective probability begin with the platitude that belief comes in degrees, in the sense that I may hold one belief more firmly than another, or that I can have varying degrees of confidence in different beliefs. The subjectivist is quick to explain, however, that he is using 'degree of belief' in a technical way. What he *means* by 'degree of belief' is something measured by betting behavior.[76] Officially, to say that a person has a degree of belief 2/3 in a proposition P is to say that he would accept a bet with 2:1 odds that P is true but would not accept a bet with less favorable odds. Given this technical construal of degrees of belief, the subjective probability of a proposition (relative to a person and a time) is identified with either (a) the person's degree of belief in that proposition, or (b) the degree of belief he rationally should have in the proposition given his overall situation. We can distinguish between these two conceptions of subjective probability as *actual degree of belief* and *rational degree of belief*.[77] It is generally claimed that subjective

---

76. See, for example, Rudolf Carnap (1962).
77. Leonard Savage (1954) proposed to distinguish between these by calling them "subjective probability" and "personalist probability", but this terminology is used only infrequently.

probability is a variety of epistemic probability.

There are problems for both conceptions of subjective probability. The principal difficulty for subjective probability as actual degree of belief is that the degrees of belief of real people will not satisfy the probability calculus. According to the probability calculus, probabilities satisfy the following three conditions:

THE PROBABILITY CALCULUS:
(1) $0 \leq \text{prob}(P) \leq 1$.
(2)   If P and Q are logically incompatible with each other then $\text{prob}(P \vee Q) = \text{prob}(P) + \text{prob}(Q)$.
(3)   If P is a tautology then $\text{prob}(P) = 1$.[78]

A person's degrees of belief are said to be *coherent* (in a sense unrelated to coherence theories of knowledge) if and only if they conform to the probability calculus. It is generally granted by all concerned that real people cannot be expected to have coherent degrees of belief. If there was ever any doubt about this, contemporary psychologists have delighted in establishing this experimentally. For some purposes the lack of coherence would not be a difficulty, but for the uses to which probability is put in epistemology it is generally essential that probability satisfy the probability calculus. Recall, for example, its use in Lehrer's theory. To carry out the kinds of calculations required by his theory, he must assume that probabilities conform to the probability calculus. It follows that subjective probability as actual degree of belief is of little use in epistemology.

Mainly because of the failure of actual degrees of belief to satisfy the probability calculus, most subjectivists adopt the "rational degree of belief" construal of subjective probability. But this construal is beset with its own problems. The first concerns whether rational degrees of belief satisfy the probability calculus any more than do actual degrees of belief. There is a standard argument that is supposed to show that they do. This is the Dutch book argument. In betting parlance, a "Dutch book" is a combination of bets on which a person will suffer a collective loss no matter what happens. For instance, suppose you are betting on a coin toss and are willing to accept odds of 2:1 that the coin will land heads and are also willing to accept odds of 2:1 that the coin will land tails. I could then place two bets with you, betting 50 cents against the coin landing heads and also betting 50 cents against the coin landing tails, with the result that no matter what happens I will have to pay you 50 cents on one bet but you will have to pay me one dollar on the other. In other words, you have a guaranteed loss—Dutch book can be made

---

78. It is customary to add a fourth axiom, to the effect that logically equivalent propositions have the same probability. This axiom implies that necessary truths have probability 1. We will not assume this axiom in the current discussion because it only exacerbates the problem of making fruitful use of probability within epistemology.

against you. The Dutch book argument consists of a mathematical proof that if your degrees of belief (which, remember, are betting quotients) do not conform to the probability calculus then Dutch book can be made against you.[79] It is alleged that it is clearly irrational to put yourself in such a position, so it cannot be rational to have incoherent degrees of belief.

A number of objections can be raised to this argument. First, there is a familiar philosophical distinction between epistemic rationality and practical rationality. Epistemic rationality is concerned with what to believe, and falls within the purview of rationality. Practical rationality is concerned with what *to do*. Practical rationality deals with prudential concerns rather than epistemic concerns. As we saw in chapter one, these are distinct concepts. The Dutch book argument seems to be concerned with practical rationality—not epistemic rationality. It may be practically irrational to put yourself in a situation in which you are guaranteed to lose, but what has that to do with the epistemic rationality of belief? This is connected with the definition of subjective probability. Subjective probability is defined to be the degree of belief it is rational to have in a proposition, but this overlooks the distinction between practical and epistemic rationality. Which should be employed in the definition of subjective probability? The degree of belief it is epistemically rational to have in a proposition looks initially like what was defined above as epistemic probability, but this does not fit well with the technical notion of 'degree of belief' defined in terms of betting behavior. It does not seem to make any sense to say that certain betting behavior is or is not epistemically rational. Only practical rationality is applicable to betting behavior, and it seems that subjective probability must be understood in this way.

Two stances are now possible. It could be that the subjectivist has simply confused these two kinds of rationality and that the confusion pervades his entire theory. A more charitable reading of subjective probability theory would take it as an explicit attempt to reduce epistemic rationality to practical rationality. On this understanding, subjective probability is defined as the degree of belief it is practically rational to have, and so understood it may be used to explicate epistemic rationality. Subjective probability might be used in different ways to explicate epistemic rationality. The simplest proposal would be to identify the epistemic probability of a proposition with its (practical) subjective probability, but other more complicated alternatives are also possible and we will say more about them in the next section.

Adopting the charitable construal of subjective probability, does the Dutch book argument establish that subjective probabilities must conform to the probability calculus? We do not think that it does. Contrary to the argument, it is not automatically irrational to accept odds allowing

---

79. This was first proven by Bruno de Finetti (1937).

Dutch book to be made against you. If you are considering a very complicated set of bets (as you would be if you were betting on all your beliefs at once), it may be far from obvious whether the odds you accept satisfy the probability calculus. If you have no reason to suspect that they do not, and could not be expected to recognize that they do not without undertaking an extensive mathematical investigation of the situation, then surely you are not being *irrational* in accepting incoherent odds. You may be making a mistake of some sort, but you are not automatically irrational just because you make a mistake.

The Dutch book argument will not do it, but perhaps there is some other way of arguing that the degrees of belief it is practically rational to have must conform to the probability calculus. Let us just pretend for the moment that this is the case. Thus one constraint on rational degrees of belief becomes satisfaction of the probability calculus. Are there any other constraints? Some probability theorists write as if this were the only constraint. But others acknowledge that there must be further constraints. For example, so-called "Bayesian epistemologists" (discussed below) adopt constraints regarding how our degrees of belief should change as we acquire new evidence. But these are still rather minimal constraints.

The question we want to raise now is whether subjective probability as *the* degree of belief one should rationally have in a proposition is well defined. Specifically, is there any reason to think that, in each specific case, there is a *unique* degree of belief it is rationally permissible to have? Consider someone who has actual degrees of belief that do not satisfy the probability calculus (as we all have). If rational degrees of belief must be unique then there must be a unique way of transforming his actual degrees of belief into ideally rational degrees of belief. Whether this is so will depend upon what constraints there are. If the only constraint is that rational degrees of belief satisfy the probability calculus, there will be infinitely many ways of changing our actual degrees of belief so that the resulting degrees of belief are rational. The coherence constraint gives us no way to choose between them, because it tells us nothing at all about how our rationally changed degrees of belief should be related to our initial degrees of belief. The coherence constraint only concerns the product of changing our degrees of belief to make them rational; it does not concern how those resulting degrees of belief are gotten from the original incoherent degrees of belief. In fact, no constraints that have ever been proposed are of any help here. Some, like the Bayesian constraints, *sound* as if they should be helpful because they concern how degrees of belief should change under various circumstances, but they are of no actual help because they assume that the degrees of belief with which we begin satisfy the probability calculus.

The possible confusion between epistemic and practical rationality is relevant here. If we understand degree of belief as degree of confidence and we are allowed to bring all the resources of epistemology to bear, it seems likely that there will be a unique degree of belief (in the sense of degree of confidence) that we should have in any particular proposition in any fixed epistemic setting. However, characterizing subjective probability in this way involves giving up its characterization in terms of practical rationality. If we proceed in this way, then it will obviously be circular to turn around and use subjective probability to analyze epistemic justification, and the latter is the avowed purpose of the externalist endeavor.

If subjective probability is to be useful to the externalist, it must be defined in terms of practical rationality rather than epistemic rationality, so we cannot appeal to epistemic constraints to guarantee that there is a unique degree of belief we rationally ought to have in each proposition. The constraints can only be practical. It might be supposed that although no one has been able to enumerate them, there are some practical constraints on rational degrees of belief that will enable us to get from incoherent actual degrees of belief to unique rational degrees of belief satisfying the probability calculus. Is this at all plausible? We think not, because we think that unlike epistemically rational degrees of confidence, rational betting quotients are not always uniquely determined by the epistemic situation. For example, suppose I hold one ticket from each of two lotteries—lottery A and lottery B. One of the lotteries is a 100-ticket lottery, and the other is a 1000-ticket lottery, but I do not know which is which. I am now required to bet on whether I will win lottery A and whether I will win lottery B. Is there a unique rational bet that I should make? It does not seem so. I know that *either* my chances of winning A are .01 and my chances of winning B are .001, *or* vice versa, but I have no way of choosing between these two alternatives. Perhaps it is irrational for me to bet in accordance with any combination of odds other than one of these two, but there can be no rational constraint favoring one of them over the other. Alternatively it might be insisted that I should regard it as equally likely that (a) lottery A has 100 tickets and lottery B has 1000, and (b) lottery A has 1000 tickets and lottery B has 100, and so I should weigh these possibilities equally and arrive at a degree of belief of $.01 \times .5 + .001 \times .5 = .0055$ for winning either lottery. But this seems wrong. Intuitively, there would be nothing irrational about betting at odds of 1:99 and 1:999 instead.[80] It certainly seems as though there is no unique rational bet in this case, and it follows that the subjective probability does not exist. Furthermore, although this is a contrived case, reflection indicates that it is not atypical of most of the situations in which we actually find ourselves. So it must be concluded that unique practically rational degrees of belief rarely, if ever, exist.

To summarize, we regard the entire theory of subjective probability

---

80. A further difficulty with the weighting strategy is that it seems to presuppose something like the Laplacian principle of insufficient reason, but as intuitive as that principle is it is also well known that it is inconsistent. In this connection, see Wesley Salmon (1966), 66ff.

as being pervasively confused, turning upon a conflation of epistemic and practical rationality. If we define subjective probability in terms of practical rationality, subjective probabilities do not exist. If instead we define subjective probability in terms of epistemic rationality and forgo the characterization of degrees of belief in terms of betting behavior, then the notion makes sense but it becomes identical with epistemic probability defined as 'degree of justification'. In the latter case, none of the technical apparatus of subjective probability theory can be brought to bear any longer. The Dutch book argument becomes inapplicable, and there is no reason to attribute any particular mathematical structure to epistemic probabilities. In fact, reasons will be given shortly for denying that epistemic probabilities satisfy the probability calculus.

This conclusion will be unpopular among externalists, because subjective probability has been the favored kind of probability for use in probabilist theories. We stand by the negative conclusions we have drawn regarding subjective probability, but a number of philosophers are too firmly wedded to subjective probability to be dissuaded by such arguments, and accordingly we will occasionally pretend that the notion makes sense in order to discuss popular versions of probabilist theories of knowledge.

2.2 Indefinite Physical Probabilities

The most popular theories of physical probability relate physical probabilities to relative frequencies. Where A and B are properties, the relative frequency, freq$[A/B]$, is the proportion of all actual B's that are A's. For example, given a coin that is tossed four times and then destroyed, if it lands heads just twice then the relative frequency of heads in tosses of that coin is $1/2$. Some theories identify the physical probability, prob(A/B), with the relative frequency freq$[A/B]$. Others take the prob(A/B) to be the limit to which freq$[A/B]$ would go if the set of all actual B's were hypothetically extended to an infinite set. Another alternative is to take prob(A/B) to be a measure of the proportion of B's that are A's in all physically possible worlds (rather than just in the actual world). On the latter theory, the connection between freq$[A/B]$ and prob(A/B) is only epistemic—observation of relative frequencies in the actual world gives us evidence for the value of the probability.[81] The physical probabilities described by all of these theories are indefinite probabilities. They relate properties rather than attaching to propositions or states of affairs.

There is something quite commonsensical about the idea that the most fundamental kind of physical probability is an indefinite probability. At the very least our epistemological access to physical probabilities is by way of observed relative frequencies, and these always concern general

---

81. Pollock's theory is of the latter sort. It is sketched in his (1984a), and worked out in detail in his (1984d) and (1989).

properties. But it cannot be denied that for many purposes we require definite probabilities rather than indefinite probabilities. This is particularly true for decision theoretic purposes. For example, if I am betting on whether Blindsight will win the third race, my bet should be based on an assessment of the probability of Blindsight winning the third race. The latter is a definite probability. It is incumbent upon any theory of physical probability to tell us how such definite probabilities are related to the more fundamental indefinite probabilities. The traditional answer has been that definite probabilities are inferred from indefinite probabilities by what is called "direct inference". The details of direct inference are problematic, and there are competing theories about how it should go, but in broad outline it is fairly simple. The basic idea is due to Hans Reichenbach (1949) who proposed that the definite probability, PROB(Aa), should be identified with the indefinite probability prob(A/B) where B includes as much information as possible about a, subject to the constraint that we have statistical information enabling us to evaluate prob(A/B). To illustrate, suppose we want to know the probability that Blindsight will win the third race. We know that he wins 1/5 of all the races in which he participates. We know many other things about him, for instance, that he is brown and his owner's name is "Anne", but if we have no information about how these are related to a horse's winning races then we will ignore them in estimating the probability of his winning this race, and we will take the latter probability to be 1/5. On the other hand, if we do have more detailed information about Blindsight for which we have statistical information, then we will base our estimate of the definite probability on that more detailed information. For example, I might know that he is injured and know that he wins only 1/100 of all races in which he participates when he is injured. In that case I will estimate the definite probability to be 1/100 rather than 1/5.

Perhaps the best way to understand what is going on in direct inference is to take the definite probability PROB(Aa) to be the indefinite probability prob($Ax/x = a$ & **K**), where **K** is the conjunction of all our justified beliefs. In direct inference we are estimating this indefinite probability on the basis of the known indefinite probability prob(Ax/Bx), where B includes all the things we are justified in believing about a and for which we know the relevant indefinite probabilities.[82]

For present purposes, the important thing to be emphasized about those definite probabilities at which we arrive by direct inference is that they are mixed physical/epistemic probabilities. We obtain definite probabilities by considering indefinite probabilities conditional on properties we are justified in believing the objects in question to possess. The epistemic element is essential here. In decision theoretic contexts

---

82. For a detailed account of direct inference based upon this idea, see Pollock's (1984d) and (1989).

we seek to take account both of the probabilistic structure of the world and of our justified beliefs about the circumstances in which we find ourselves.

It is worth noting that an analogous account of direct inference to non-epistemic definite probabilities cannot work. Such an account would have us estimating the definite probability of Aa by considering all *truths* about a rather than all justified beliefs about a. In other words, in direct inference we would be trying to ascertain the value of $\text{prob}(Ax/x = a \,\&\, T)$ where T is the conjunction of all truths. But among the truths in T will be either Aa or $\sim\!Aa$, so $\text{prob}(Ax/x = a \,\&\, T)$ is always either 1 or 0 depending upon whether Aa is true or false. Direct inference to such non-epistemic probabilities could never lead to intermediate values. Such an account of definite probabilities and direct inference would be epistemologically useless.

2.3 Definite Propensities

An approach to physical probabilities that is less popular but by no means moribund takes the fundamental physical probabilities to be definite probabilities that pertain to specific individuals. These are propensities. We can, for example, talk about the propensity of a particular coin to land heads on the next toss. These are supposed to be purely physical probabilities, untinged by any epistemic element, and are supposed to reflect ineliminable chance relationships in the world.[83] The defenders of propensities usually agree that propensities would always be either 0 or 1 in a deterministic world.[84]

Propensity theories have not been developed to the same extent as subjective theories and frequency theories, and most philosophers remain suspicious of propensities, but it would be premature to reject them outright. We must at least keep them in the back of our minds in considering what kinds of probabilities to use in formulating externalist theories of knowledge.

3. Probabilism

The distinction between probabilism and reliabilism can now be made precise by saying that probabilism seeks to characterize epistemic justification in terms of the definite probabilities of one's beliefs, while reliabilism seeks to characterize epistemic justification in terms of more general indefinite probabilities pertaining to such things as the reliability

---

83. Propensity theories have been proposed by Ian Hacking (1965), Isaac Levi (1967), Ronald Giere (1973), (1973a), and (1976), James Fetzer (1971), (1977), and (1981), D. H. Mellor (1969) and (1971), and Patrick Suppes (1973). A good general discussion of propensity theories can be found in Ellory Eells (1983).
84. Ronald Giere (1973), p. 475.

of the cognitive processes that produce the beliefs. Probabilism represents the most straightforward way of trying to capture the intuition that in acquiring beliefs we should adopt only probable beliefs.

3.1 The Simple Rule and Bayesian Epistemology

The simplest form of probabilism endorses what we call *the simple rule*:

A person is justified in believing *P* if and only if the probability of *P* is sufficiently high.

This rule seems intuitively quite compelling, and at various times it has been endorsed by a wide spectrum of philosophers.[85] The simple rule, if acceptable, admirably satisfies the requirement that an analysis of epistemic justification must explain why the notion should be of interest to us. What could be more intuitive than the claim that in deciding what to believe we should be trying to ensure that our beliefs are probably true? The endorsement of the simple rule allows us to bring all of the mathematical power of the probability calculus to bear on epistemology, and the results have often seemed extremely fruitful. The simple rule pertains most directly to what we believe in a fixed epistemic setting, but what happens when our epistemic situation changes through the acquisition of new data (e.g., in perception)? Probabilists typically appeal to what is known as Bayes' Theorem,[86] according to which

$$\text{prob}(P/Q) = \text{prob}(Q/P) \times \frac{\text{prob}(P)}{\text{prob}(Q)} .$$

Taking *P* to be the proposition whose epistemic status is to be evaluated and *Q* to be the new evidence, prob(*P*/*Q*) is interpreted as the probability of *P* given the new evidence; prob(*P*) is the prior probability of *P* (i.e., the probability prior to acquiring the evidence); prob(*Q*) is the prior probability of acquiring that evidence; and prob(*Q*/*P*) is the prior probability of acquiring the evidence given the specific assumption that *P* is true. This principle then tells us how to alter our probability assignments in the face of new evidence.

Epistemology based upon the simple rule and Bayes' Theorem is known as *Bayesian epistemology*.[87] It has exerted a strong influence on technically minded philosophers, partly because of the intuitiveness of its basic principles and partly because of its mathematical elegance and

---

85. It was endorsed, for example, by Roderick Chisholm (1957), p. 28, and Carl Hempel (1962), p. 155. Its most ardent recent defender is probably Henry Kyburg (1970) and (1974). See also Richard Jeffrey (1970), Rudolf Carnap (1962) and (1971), David Lewis (1980), and Isaac Levi (1980).

86. After Thomas Bayes.

87. Sometimes the term 'Bayesian epistemology' is reserved for theories proceeding in terms of subjective probability.

power. It has spawned an extensive literature and has seemed to be extremely fruitful when applied to problems like the problem of induction or the analysis of the confirmation of scientific theories. Unfortunately, Bayesian epistemologists have concentrated more on the mathematical elaboration of the theory than on its foundations. There are major problems with the foundations. These concern the very idea of acquiring new data. Note that it follows from the proposed interpretation of Bayes' theorem that when we acquire new data Q, it will come to have probability 1. This is because $\text{prob}(Q/Q) = 1$. But what is it to acquire new data through, for instance, perception? This is just the old problem of accommodating perceptual input within an epistemological theory. The beliefs we acquire through perception are ordinary beliefs about physical objects, and it seems most unreasonable to regard them as having probability 1. Furthermore, it follows from the probability calculus that if $\text{prob}(Q) = 1$ then for any proposition R, $\text{prob}(Q/R) = 1$. Thus if perceptual beliefs are given probability 1, the acquisition of further data can never lower that probability. But this is totally unreasonable. We can discover later that some of our perceptual beliefs are wrong.[88]

The idea that "data" should receive probability 1 is reminiscent of the appeal to epistemologically basic beliefs. What is happening here is that although Bayesian epistemology is a nondoxastic theory, it is nondoxastic in the wrong way. Doxastic theories fail to handle perception correctly because the only internal states to which they can appeal are beliefs. Specifically, they do not appeal to perceptual states. But beliefs are also the only internal states to which Bayesian epistemology appeals. Bayesian epistemology is nondoxastic because it appeals to probability, not because it appeals to internal states other than beliefs. Consequently, Bayesian epistemology encounters precisely the same sort of problem as do doxastic theories in accommodating perception. Contrary to both doxastic theories and Bayesian epistemology, the justifiability of a perceptual belief is partly a function of nondoxastic internal states.

It should be emphasized that the use of Bayes' theorem in describing perception and belief change is not required by the simple rule. These are independent principles. Thus we can reject Bayesian epistemology without rejecting the simple rule. The simple rule might be combined with a more sophisticated account of perception and memory without robbing it of the power inherent in its use of probability. And the simple rule would still retain its intuitive appeal in capturing the idea that what we should be doing in the epistemological evaluation of beliefs is choosing beliefs that are probable. Unfortunately, this elegant rationale for the simple rule begins to crumble when we examine the rule more closely. In evaluating the simple rule we must decide what kind of probability is

---

88. Richard Jeffrey (1965) has proposed a variant of the Bayesian rule that avoids this problem by allowing new data to have probability less than 1, but it does not tell us how to assign probability to the new data.

involved in it. It is a definite probability, but we have taken note of the (at least putative) existence of four distinct kinds of definite probability: epistemic probability, subjective probability, mixed physical/epistemic probability, and propensities.

3.1.1 Epistemic Probabilities

The simple rule is a truism when interpreted in terms of epistemic probabilities. So understood it claims no more than that a belief is justified if and only if its degree of justification is sufficiently high. This claim cannot be faulted as long as it is understood that there is no presupposition either that epistemic probabilities are quantifiable or that if they are quantifiable then they satisfy the probability calculus. But, of course, understood in this way the simple rule is trivial and unilluminating. In particular, it does not constitute an analysis of epistemic justification, because epistemic probability is itself defined in terms of epistemic justification.

3.1.2 Subjective Probabilities

What is no doubt the favorite interpretation of the simple rule in contemporary philosophy is the one proceeding in terms of subjective probability. So construed, the simple rule becomes the claim that epistemic probabilities are the same as subjective probabilities. We have argued that no sense can be made of subjective probability, and we regard this as the most serious objection to the endorsement of the simple rule construed in terms of subjective probability. But suppose we waive this difficulty, pretending that there is always a unique betting quotient that one practically should accept for each given proposition. This construal of the simple rule then yields an analysis of epistemic justification that is not obviously circular, and an immense literature has grown up around it. Is this a plausible analysis?

The simple rule, construed in terms of subjective probabilities as actual degrees of belief, would tell us that a belief is justified if and only if it is firmly held. That obviously has nothing to recommend it, so let us confine our attention to subjective probabilities as rational degrees of belief. A simple objection to the use of such subjective probabilities in the simple rule consists of questioning whether practical rationality can be understood without first understanding epistemic rationality. It certainly seems that what we practically should do is a function in part of what we (epistemically) reasonably believe. Whether I should bet that Blindsight will win the next race is going to be determined in part by what I am justified in believing about Blindsight. If it is correct that practical rationality presupposes epistemic rationality, then it becomes circular to analyze epistemic rationality in terms of practical rationality, and hence it becomes circular to analyze epistemic rationality in terms of subjective probabilities.

3.1.3 Mixed Physical/Epistemic Probabilities

We might interpret the simple rule in terms of the mixed physical/epistemic probabilities that are obtained from indefinite physical probabilities by direct inference. This is the proposal of Henry Kyburg.[89] But any such proposal is subject to a simple objection—the resulting analysis of epistemic justification is circular. Recall that these physical/epistemic definite probabilities are obtained by direct inference from indefinite probabilities, and direct inference proceeds by considering indefinite probabilities conditional on *what we are justified in believing* about the objects in question. For example, what makes it true that the probability (for me) is 1/2 of Jamie having an accident while driving home this evening is that I am *justified in believing* that he is inebriated, driving on busy streets, and so on, and the indefinite probability of someone having an accident under those circumstances is 1/2. It is not the mere *fact* that Jamie is inebriated that makes the probability high. Only what I am justified in believing about Jamie can affect the mixed physical/epistemic probability. Thus mixed physical/epistemic probabilities cannot be used non-circularly in the analysis of epistemic justification.

3.1.4 Propensities

An analysis of epistemic justification in terms of propensities would not be obviously circular. Can we take a proposition to be justified if and only if it has a sufficiently high propensity to be true? It is hard to evaluate this suggestion without a better understanding of propensities. Notice, however, that according to most propensity theorists, nontrivial propensities only exist in nondeterministic worlds. If the world were deterministic, all propensities would be either 0 or 1, depending upon whether the proposition in question were true or false. The simple rule would then reduce to the absurd principle that we are justified in believing something if and only if it is true. We might avoid this objection by insisting that nontrivial propensities exist even in deterministic worlds, but to defend this we need a better understanding of propensities than anyone has yet provided. It is rather difficult to say much specifically about the propensity interpretation of the simple rule without a better theory of propensities.

3.1.5 General Difficulties

We have raised objections to each of the interpretations of the simple rule in terms of different kinds of definite probabilities. Some of those objections are more telling than others. We turn now to some more general objections that apply simultaneously to all versions of the simple rule.

---

89. In Kyburg (1974), and elsewhere.

(a) Tautologies

A reasonably familiar objection is that it follows from the probability calculus that every tautology has probability 1.[90] It would then follow from the simple rule that we are justified in believing every tautology. Such a conclusion is clearly wrong. If we consider some even moderately complicated tautology such as

$$[P \leftrightarrow (R \vee \sim P)] \rightarrow R$$

it seems clear that *until we realize that it is a tautology*, we are not automatically justified in believing it. The only way to avoid this kind of counterexample to the simple rule is to reject the probability calculus, but that is a very fundamental feature of our concept of probability and rejecting it would largely emasculate probability. It is because probability has the nice mathematical structure captured by the probability calculus that it has proven so fruitful, and that mathematical structure has played an indispensable role in the employment of probability in epistemology.

(b) Epistemic Indifference

The preceding difficulty illustrates one respect in which epistemic justification seems to have a more complicated structure than can be captured by the probability calculus. Supposing that epistemic probability satisfies the probability calculus forces us to regard different propositions (different tautologies) as equally justified when it seems clear that we want to make epistemic distinctions between them. The converse problem also arises. There are cases in which we want to regard propositions as equally justified when the probability calculus would preclude that. Consider a pair of unrelated propositions, *P* and *Q*, regarding which we know essentially nothing. Under the circumstances, we should neither believe these propositions nor disbelieve them—we should withhold belief. We can express this by saying that we should be *epistemically indifferent* with respect to *P*, and also with respect to *Q*. We are precisely as justified (or unjustified) in believing *P* and in believing *~P*, and similarly for *Q*. The simple rule and the probability calculus would require that in this sort of case, prob(*P*) = prob(*~P*) = 1/2, and prob(*Q*) = prob(*~Q*) = 1/2. But now consider the disjunction (*P* ∨ *Q*). If we are completely ignorant regarding *P* and *Q* and they are logically unrelated, then we are

---

90. We have avoided endorsing the standard axiom requiring logically equivalent propositions to have the same probability. That axiom must be added for a reasonable axiomatization of probability, but we have not endorsed it here in order to avoid begging questions against probabilism. Probabilism would encounter even more severe difficulties in the face of that axiom and its consequence that all necessary truths (not just tautologies) have probability 1.

also completely ignorant regarding $(P \vee Q)$. We should withhold belief with respect to $(P \vee Q)$ just as we did with respect to P and with respect to Q. Thus we should conclude, as above, that prob$(P \vee Q) = $ prob$(\sim(P \vee Q)) = 1/2$. The difficulty is that the probability calculus commits us to regarding $(P \vee Q)$ as more probable than either P or Q individually, so we cannot assign probability $1/2$ to all three of P, Q, and $(P \vee Q)$.[91] Thus the simple rule precludes our being epistemically indifferent to all three of P, Q, and $(P \vee Q)$, and yet intuition seems to indicate that there are circumstances under which such indifference is epistemically prescribed. Once again, the structure of epistemic justification is not properly reflected by the probability calculus.

(c) The Simple Rule Is Self-Defeating
   The preceding objections to the simple rule are familiar ones in the philosophical literature, but for reasons that escape us, they have failed to make many confirmed probabilists repent. So consider a third objection, which seems to us show that probabilism based on the simple rule is a hopeless theory. This third objection argues that the theory is "self-defeating", in the sense that if it were true it would make it impossible for us to have any interesting justified beliefs. The argument is as follows. According to the simple rule, to be justified a belief must be highly probable. Many of the beliefs that we regard as justified are obtained by inference from other justified beliefs. We would like to be able to apply logical inference "blindly", simply assuming that if a conclusion is deduced from other highly probable conclusions, then it is itself highly probable. For this to be possible, the inference must proceed in terms of a "probabilistically valid" inference rule, where this notion is defined as follows:

DEFINITION:
An inference rule is *probabilistically valid* if and only if it follows from the probability calculus that whenever a conclusion can be inferred in accordance with it from a set of premises, the probability of the conclusion is at least as great as the probability of the least probable premise.

It is a consequence of the probability calculus that if P logically entails Q then PROB$(Q) \geq$ PROB(P). So deductive inference from a single premise always preserves high probability, and such inferences are probabilistically valid. This seems to lend credence to the claim that inferences should always be made in accordance with probabilistically valid inference rules. However, many important inference rules proceed from multiple premises. Here are two such rules:

---

91. Technically, prob$(P \vee Q) = $ prob$(P) +$ prob$(Q) -$ prob$(P$ & $Q)$, and because P and Q are unrelated, prob$(P$ & $Q) = $ prob$(P) \times$ prob$(Q) = 1/4$, with the result that prob$(P \vee Q) = 3/4$.

ADJUNCTION   MODUS PONENS

$$\frac{P, Q}{(P \& Q)}$$

$$\frac{P, (P \to Q)}{Q}$$

A problem now arises. Neither of these rules is probabilistically valid. The probability calculus implies that PROB(P & Q) ≤ PROB(P) and PROB(P & Q) ≤ PROB(Q). If P and Q are independent, PROB(P & Q) = PROB(P) × PROB(Q), and so unless PROB(P) = 1 or PROB(Q) = 1, PROB($P\&Q$) will be less than either PROB(P) or PROB(Q). So adjunction is not probabilistically valid. The situation is similar for modus ponens. Nor is this just a problem for these two inference rules. Let us say that a premise *occurs essentially* in an inference rule if the rule becomes invalid when we delete the premise. It then turns out that *no* deductive inference rule is probabilistically valid if it has multiple premises all of which occur essentially. Thus the probabilist is committed to asserting that no such inference rules can be used blindly in drawing new justified conclusions. For instance, if a person justifiably believes P and justifiably believes Q, she cannot automatically infer (P & Q). Instead, she must somehow compute the probability of (P & Q) and then decide on that basis whether to believe it.

This is extremely counterintuitive. For instance, consider an engineer who is designing a bridge. She will combine a vast amount of information about material strength, weather conditions, maximum load, costs of various construction techniques, and so forth, to compute the size a particular girder must be. These various bits of information are, presumably, independent of one another, so if the engineer combines 100 pieces of information, each with a probability of .99, the conjunction of that information has a probability of only $.99^{100}$, which is approximately .366. According to the probabilist, she would be precluded from using all of this information simultaneously in an inference—but then it would be impossible to build bridges.

As a description of human reasoning, this seems clearly wrong. Once one has arrived at a set of conclusions, one does not hesitate to make further deductive inferences from them. But an even more serious difficulty for the probabilist is that the simple rule turns out to be self-defeating; if it were correct, it would be impossible to perform the very calculations required by the simple rule for determining whether a belief ought to be held. This arises from the fact that the simple rule would require a person to decide what to believe by computing probabilities. The difficulty is that the probability calculations themselves cannot be performed by a probabilist who endorses the simple rule. To illustrate the difficulty, suppose the reasoner has the following beliefs:

PROB($P \lor Q$) = PROB(P) + PROB(Q) − PROB(P & Q)
PROB(P) = .5
PROB(Q) = .49
PROB(P & Q) = 0.

From this we would like the reasoner to compute that PROB$(P \vee Q)$ = .99, and perhaps go on to adopt $(P \vee Q)$ as one of her beliefs. However, the probabilist cannot do this. The difficulty is that this computation is an example of a "blind use" of a deductive inference, and as such it is legitimate only if the inference is probabilistically valid. To determine the probabilistic validity of this inference, it must be treated on a par with all the other inferences performed by the reasoner. Although the premises are about probabilities, they must also be assigned probabilities ("higher-order probabilities") to be used in their manipulation. Viewing the inference in this way, we find that although the four premises do logically entail the conclusion, the inference is not probabilistically valid for the same reason that *modus ponens* and adjunction fail to be probabilistically valid. It is an inference from a mutliple premise set, and despite the entailment, the conclusion can be less probable than any of the premises.

The upshot of this is that the simple rule is self-defeating. On the one hand, it precludes a person from drawing conclusions on the basis of probabilistically invalid deductive inferences, requiring one instead to decide whether to believe putative conclusions by computing their probabilities. But on the other hand it makes it impossible to compute those probabilities because the computations involved are probabilistically invalid. The inescapable conclusion is that the simple rule is bankrupt.

We have raised three "formal" objections to the simple rule. It should be pointed out that these objections to the simple rule are also objections to a more general principle. A number of philosophers reject the identification of epistemic justification with probability, but nevertheless maintain that degree of epistemic justification "work like" probabilities, in the sense that they satisfy the probability calculus.[92] That claim is made equally untenable by the formal objections.

We regard these objections to the simple rule as decisive. The structure of epistemic justification is too complicated to be captured by the probability calculus. Why then does the rule seem so intuitive? We think that there is a twofold explanation for this. First, a very common use of 'probable' in English is to express epistemic probability, and the simple rule understood in terms of epistemic probability is a truism. Philosophers confuse this truism with more substantive versions of the simple rule that proceed in terms of other varieties of probability. Only those more substantive versions can hope to provide a noncircular account of epistemic justification, but those more substantive versions have fatal flaws and are not themselves directly supported by our intuitions.

The second part of the explanation amounts to observing that we also have the intuition that we should not believe something if it is improbable, where the kind of probability involved is the mixed physical/epistemic probability involved in decision theory. This intuition is correct, but it lends no support to probabilism. It is correct because mixed physical/epistemic probabilities are conditional on the conjunction of all justi-

---

92. Richard Fumerton (1995).

fied beliefs, and hence the probability of any justified belief will automatically be 1.[93] But this lends no support to probabilism, because mixed physical/epistemic probabilities are defined in terms of justified belief and hence cannot be used noncircularly to analyze epistemic justification.

3.2 Other Forms of Probabilism

The simple rule is the most natural form of probabilism, but it fails due to the fact that the mathematical structure of epistemic justification cannot be captured by the probability calculus and hence degree of justification cannot be identified with any kind of probability conforming to the probability calculus. However, it is possible to construct more sophisticated forms of probabilism that escape this objection. These theories characterize epistemic justification in terms of probabilities, but they do not simply identify degree of justification with degree of probability. For example, recall Keith Lehrer's coherence theory. Its central thesis is:

P is justified for S if and only if for each proposition Q competing with P, S believes P to be more probable than Q.

This is a doxastic theory because its appeal to probability is only via beliefs about probability, but it could be converted into a nondoxastic theory by appealing to the probabilities themselves:

P is justified for S if and only if P is more probable than each proposition competing with P.

Supplementing this central criterion with a definition of competition will yield a version of probabilism more complicated than the simple rule. We might adopt Lehrer's definition of competition, or we might adopt another definition. Marshall Swain (1981) follows essentially this course. A slightly simplified version of his definition of competition is as follows:

Q is a competitor of P for S if and only if either:
(1) (a) P and Q are contingent,
    (b) Q is negatively relevant to P, and
    (c) Q is not equivalent to a disjunction of propositions one of whose disjuncts, R, is both (i) irrelevant to P and (ii) such that the probability of R is greater than or equal to the probability of P; or
(2) P is noncontingent and Q is ~P (p. 133).

Swain's theory is just one example of a sophisticated kind of probabilism that escapes the objection to the simple rule that the structure of epistemic justification is too complicated to be captured by the probability calculus. Thus the formal objections to the simple rule do not refute probabilism in general.

---

93. It is a theorem of the probability calculus that $prob(P/P \& Q) = 1$.

Still, two general points can be made about all versions of probabilism. First, there appears to be no appropriate kind of probability for use in probabilist theories of knowledge. Such theories require a definite probability. They are circular if formulated in terms of epistemic probability. Only three other kinds of definite probability have been discussed in the literature: subjective probability, propensities, and mixed physical/epistemic probability. We have argued that the very concept of subjective probability is ill-defined—subjective probabilities do not exist. We regard mixed physical/epistemic probabilities as unproblematic, but they cannot be used in the analysis of epistemic justification because they already presuppose epistemic justification. Propensities are not sufficiently well understood to be of much use anywhere. Furthermore, a common view among propensity theorists is that nontrivial propensities only exist in nondeterministic worlds, but it seems pretty clear that any probabilist analysis of epistemic probability must proceed in terms of a variety of probability that can take values intermediate between 0 and 1 even in deterministic worlds. It seems inescapable that there is no appropriate kind of probability for use in probabilist theories of knowledge.

This point applies to any theory that has a probabilistic component embedded in it, no matter what other elaborations are proposed. For instance, William Alston (1988) attempts to combine elements of internalism and externalism by defending a view that has an internal access constraint while at the same time demanding that the grounds for a belief make it probable that the belief is true. Most of his efforts have been in developing the internalist side of his view. He acknowledges that it is extremely difficult to state with any precision just how the probabilistic part of the theory should be developed. We have offered some indication of why that is, and have pointed out some reasons for being pessimistic.

The second point to be made about probabilist theories concerns not their truth but their motivation. The original motivation for probabilism came from the intuition that what the epistemic evaluation of beliefs should be trying to ensure is that our beliefs are probable. If the simple rule were defensible, it would capture that intuition. But the simple rule is not defensible, and more complicated versions of probabilism do not capture this intuition. In fact, they are incompatible with the intuition—insofar as they diverge from the simple rule, they will have the consequence that we can be justified in believing improbable propositions and unjustified in believing probable propositions.

The intuition that we should only believe things when they are probable is a powerful one. It seems to lead directly to the simple rule, but there are overwhelming objections to the simple rule. What, then, should we make of this intuition? As we have indicated, it is epistemic probability that is involved in this intuition. In ordinary non-philosophical English, "probable" is used to express epistemic probability at least as often as it is used to express other kinds of probability, and it is a truism that a belief is justified if and only if its epistemic probability is sufficiently high. But, of course, epistemic probability is defined in terms of epistemic justification, so this provides no analysis of epistemic justification and no support for probabilism.

What should we conclude about probabilism at this point? Decisive objections can be raised against existing probabilist theories of knowledge, and they militate strongly against there being any defensible kind of probabilism. At the very least, the probabilist owes us an account of the kind of probability in terms of which he wants his theory to be understood, and there is good reason for being skeptical about there being any appropriate kind of probability. The most rational attitude to adopt towards probabilism at this point is healthy skepticism, but it cannot be regarded as absolutely certain that no probabilist theory can succeed. In chapter five, we will eventually present arguments that purport to show that no externalist theory of any kind can be correct, and those arguments should lay probabilism to rest for good.

4. Reliabilism

Reliabilism differs from probabilism in that it attempts to analyze epistemic justification in terms of indefinite probabilities rather than definite probabilities. Existing reliabilist theories do this by appealing to the reliability of cognitive processes. We call this process reliabilism. The basic idea behind process reliabilism is that a belief is justified if and only if it is produced by a reliable cognitive process.[94] For example, the reliabilist explains why perceptual beliefs are justified by pointing to the fact that, in the actual world, perception is a reliable cognitive process. Similarly, deduction is a reliable cognitive process, so beliefs deduced from other justified beliefs will be justified. On the other hand, wishful thinking is not a reliable cognitive process, so beliefs produced by wishful thinking are not justified. The reliability of a cognitive process is the indefinite probability of beliefs produced by it being true. This is a perfectly respectable notion. Unlike probabilism, we cannot fault process reliabilism for making illegitimate use of probability.

Let us consider then just how a reliabilist theory might be formulated. To this end, we will give a sketch of Alvin Goldman's theory, as presented in his article "What is justified belief?"

---

94. Reliabilist theories of *knowing* predate reliabilist theories of justification. The first such theories were those of Frank Ramsey, David Armstrong, and W. V. O. Quine. More recently, Fred Dretske has defended a reliabilist theory of knowledge (1981). In this book, we will confine the term 'reliabilism' to reliabilist theories of epistemic justification. Perhaps the first formulation of a reliabilist theory of justification is due to Wilfrid Sellars (1963), but he formulated it only to reject it. Current forms of reliabilist theories of justification take their impetus from the work of Alvin Goldman. See his (1979), (1981), and (1986). Goldman's (1988), (1992), and (1994) also represent significant developments in his thinking about reliabilism.

4.1 Goldman's Theory

Goldman begins by distinguishing between two kinds of cognitive processes—*belief-dependent* processes such as reasoning that take beliefs as inputs, and *belief-independent* processes such as perception that do not. Focusing first on belief-independent processes, Goldman writes:

> Consider some faulty process of belief-formation, i.e., processes whose belief-outputs would be classed as unjustified. Here are some examples: confused reasoning, wishful thinking, reliance on emotional attachment, mere hunch or guesswork, and hasty generalization. What do these faulty processes have in common? They share the feature of *unreliability*: They tend to produce *error* a large proportion of the time. By contrast, which species of belief-forming (or belief-sustaining) processes are intuitively justification-conferring? They include standard perceptual processes, remembering, good reasoning, and introspection. What these processes seem to have in common is *reliability*: the beliefs they produce are generally reliable (1979, 9-10).

In light of this diagnosis, Goldman proposes the following:

If S's belief in P at t results ("immediately") from a belief-independent process that is reliable, then S's belief in P at t is justified.

Turning to belief-dependent processes, Goldman defines such a process to be *conditionally reliable* if and only if it tends to produce true beliefs when the input beliefs are true. For example, deduction and induction are both conditionally reliable belief-dependent processes. Goldman then makes the provisional proposal:

> If S's belief in P at t results ("immediately") from a belief-dependent process that is (at least) conditionally reliable, and if the beliefs (if any) on which this process operates in producing S's belief in P at t are themselves justified, then S's belief in P at t is justified.

This proposal is only provisional because Goldman goes on to recognize that we must take account of what is in effect defeasibility (although he does not call it that). We might arrive at a belief by employing a relatively simple cognitive process (e.g., color perception), but that belief would be unjustified if there were a more elaborate reliable cognitive process (one taking account of additional information, such as abnormal lighting conditions) such that had we employed the more elaborate process we would not have adopted the belief. This leads Goldman to replace the provisional proposal with the following:

> If S's belief in P at t results from a reliable cognitive process, and there is no reliable or conditionally reliable process available to S which, had it been used by S in addition to the process actually used, would have resulted in S's not believing P, then S's belief in P at t is justified.

Goldman's reliability theory generates a structure of justified beliefs

having the same kind of pyramid structure that is embodied in a foundations theory or in direct realism.[95] Perceptual input, by being reliable,
produces beliefs that are justified simply by virtue of being produced in
that way. Subsequent reasoning then produces further beliefs that are
justified by being produced by conditionally reliable cognitive processes
that are applied to beliefs already justified. But the theory diverges
radically from both foundations theories and direct realism in that it
makes the justifiedness of a belief depend not only on the processes that
produced it, but also on whether those processes happen to be reliable in
the actual world. By contrast, internalist theories would rule that if a
particular combination of perceptual inputs, reasoning, and so on, produces justified belief in the actual world then it will produce justified
belief in all possible worlds. Internalist theories preclude appeal to external
considerations such as reliability in evaluating beliefs.

4.2 Problems for Process Reliabilism
Several different kinds of problems can be generated for reliabilism.
The simplest is that, intuitively, reliability has nothing to do with epistemic
justification. We would certainly expect those cognitive processes producing justified beliefs to be reasonably reliable, because if they were not
the human race would have vanished long ago, but it is not the reliability
of the processes that is responsible for the justifiedness of the beliefs.
The beliefs are justified just because the believer is "reasoning correctly"
(in a broad sense of "reasoning"). If one makes all the right epistemic
moves, then one is justified regardless of whether her belief is false or
nature conspires to make such reasoning unreliable. To us, this objection
seems to have considerable intuitive force, but reliabilists deny the intuition. Internalists have tried to make this objection more precise by
arguing that reliabilist theories give assessments of justifiedness that are
intuitively wrong in some cases. A favorite example is that of a brain in
a vat. Recall poor Harry, whom we met in chapter one. He had his
brain removed and wired into a computer that directly stimulated his
visual cortex so that he had normal-seeming sensory experiences but
they were totally unrelated to his physical surroundings. For Harry,
perception became an unreliable cognitive process, and thus the reliabilist
is committed to regarding Harry's perceptual beliefs as unjustified. But

---

95. Some philosophers treat reliabilist theories as a species of
foundationalism, although in doing so they are focusing on just the structural aspects
of foundationalism and detaching those from the endorsement of the doxastic
assumption. See Ernest Sosa (1980) and William Alston (1976). We are inclined to see
this as a overly restrictive conception of reliabilism, though it does point to potentially
interesting interactions between characteristically internalist discussions of belief
structure and externalist considerations. Along these lines, see Laurence Bonjour (1985)
where he formulates an externalist version of coherentism (though only to reject it).
Goldman offers an extended discussion of how the foundationalist or coherentist
structure is compatible with reliabilism in his (1986), chapter four.

this seems wrong. Harry has no reason to suspect that anything is amiss, so if he takes reasonable care in forming perceptual judgments we will regard them as justified.

Another well-known case that is designed to show the inadequacy of reliabilism is due to Laurence Bonjour (1980), p. 62 and (1985), p. 41. He invites us to consider Norman, who has reliable clairvoyance and who believes that he has clairvoyance, although he lacks any justification for that belief. One day Norman comes to believe that the President is in New York City on the basis of his clairvoyance. Bonjour claims that our intuitive response is that Norman, in the absence of any justification for his belief in his own clairvoyance, is irrational to accept his belief that the President is in New York. Since reliabilism claims that Norman is justified in that belief, it gives the intuitively wrong results.[96]

We have distinguished between two different concepts of epistemic justification—procedural epistemic justification, and the "knowledgifying" concept, i.e., the kind of justification that is supposed to turn true belief into knowledge. One might reasonably suspect that the conflicting intuitions generated by examples of brains in vats and Norman-type examples just reflect the fact that different philosophers are talking about different concepts of justification. It seems initially plausible that, for instance, Harry might hold procedurally justified beliefs, but lack knowledge even in those cases in which his procedurally justified beliefs happen more or less by chance to be true. However, our experience has been that reliabilists are not willing to concede that this is the source of the dispute.

At this point, the debate between the internalists and the reliabilists often reduces to a shoving match of conflicting intuitions, each insisting that *his* intuitions support *his* theory. Such a conflict of intuitions is hard to resolve. There is, however, an argument due to Stewart Cohen (1984) that appears to weigh heavily in favor of the internalist. Cohen notes that even in the case of a brain in a vat, we would distinguish between reasonable and unreasonable epistemic behavior. To illustrate, we noted earlier that perceptual beliefs may be unjustified when the perceiver fails to take account of features of the perceptual situation that he knows or believes to adversely affect the reliability of perception. Consider a person who knows all about the way in which colored lights can affect apparent color. He will be unjustified in judging color on the basis of apparent color when he believes that what he sees is bathed in colored lights. Harry presumably knows all about such phenomena (from before his brain was removed from his skull and installed in the vat), so if he judges something to be red under circumstances that appear normal to him he will be justified, but if he makes the same judgment under circumstances in which he thinks the object is illuminated by red lights then he will be unjustified. We would insist upon making such epistemic

---

96. Also see Alvin Plantinga (1988) and (1993a), and Keith Lehrer (1990) for counterexamples to reliabilism.

discriminations despite the fact that his perceptual judgments are uniformly unreliable.

Goldman (1986) formulates a reliabilist theory that is explicitly intended to avoid brain-in-a-vat and evil demon cases. He defines a normal world to be one consistent with our *general* beliefs about the actual world, and then proposes that justification requires production by cognitive processes that are reliable in normal worlds (but not necessarily in the actual world). A simple objection to this theory is that it puts no constraints on how we get our general beliefs. If they are unjustified, then it seems that reliability relative to them should be of no particular epistemic value. Goldman himself (1986, p. 102) raises basically the same objection to epistemic decision theory based upon subjective probabilities. To paraphrase that objection and apply it to Goldman's own theory:

> Apparently the [general beliefs] may be formed in any fashion at all, including hunch, fancy and the like. But if a belief is based on [a process reliable only relative to general beliefs] of that ilk, there are no grounds for regarding the belief as justified. It certainly is not justified in any sense that links up closely with knowledge.

Note in particular that if our general beliefs are not formed by processes reliable in the actual world, then there is no reason to regard beliefs formed by processes reliable relative to those general beliefs as being probably true. Thus this strategy for avoiding brain-in-a-vat counterexamples appears to forsake the general intuitions that made reliabilism attractive in the first place.

The same difficulties apply to Goldman's most recent version of a reliabilist theory (1992). In that account, a belief is justified if it is judged to be the product of a chain of "virtuous" psychological processes. A belief is unjustified if it is judged to be partially or wholly the product of a chain of "vicious" psychological processes. On his proposal, the judge or epistemic evaluator possesses a list of which processes are which. This list encodes our concept of justification, and we judge novel cases according to the similarity between novel cases and our psychological examplar of a virtuous cognitive process. If the processes implicated in the novel case do not appear on either the virtuous process or vicious process list, then the belief that results from them is non-justified. Goldman intends this theory to be broadly consistent with his other views. The account is externalist since justification is determined by whether or not those who are epistemically evaluating us deem the processes that we use virtuous. The account remains reliabilist in that reliability is the deeper standard by which people or communities judge inclusion on the virtue list.

Goldman admits that this proposal is properly a description of our epistemic judgments and is more suited to explaining why we make the epistemic judgments that we do rather than explaining what epistemic achievement is. The sense of reliability that motivates people's psychological concept of justification must be people's beliefs about reliability rather than actual reliability. This is because the psychological exemplars could have no way encoding actual reliability. But, again, it is not clear why we should place any special value on people's beliefs about reliability. Goldman's latest theory seems to represent a turning away from the project of giving a philosophical analysis of justification. Rather, he gives a psychological analysis of the conditions under which we are inclined to judge cognitive processes as virtuous or vicious. This might be an appropriate response to a complete failure on the part of epistemologists to illuminate normative procedural epistemology. Goldman, however, has not demonstrated that complete failure.

We turn now to a way of attacking process reliabilism that raises very fundamental difficulties for the theory. We have taken it for granted that our ordinary cognitive processes such as color vision are reliable, but are they? Color vision is reliable under some circumstances (e.g., standard lighting conditions) and unreliable under other circumstances (e.g., illumination by colored lights). Judgments of reliability are usually made relative to rather narrowly circumscribed circumstances. We *can* talk about reliability in the universe at large, but the lighting conditions on earth are quite unusual by universe-wide standards. In the universe at large, color vision is unreliable. As that does not incline us to regard our normal color judgments as unjustified, it must be concluded that it is not reliability in the universe at large that is relevant. It must instead be reliability in the circumstances in which we actually find ourselves. But what does this mean? We find ourselves in many circumstances. We find ourselves in the universe at large, and we also find ourselves on earth viewing things in ordinary daylight. Relative to which set of circumstances are we to judge reliability? Obviously, the latter, but why? The intuitive answer is "Because it takes account of more relevant information."

It might first be proposed that in judging the reliability of a cognitive process in a particular instance we should take account of *everything* about the circumstances in which it is used. If it makes any sense at all to talk about the reliability of the cognitive process "under the present circumstances" (in all their specificity), it seems that it must be the indefinite probability of producing a true belief, conditional on everything true of the present circumstances. But the present circumstances are infinitely specific and include, among other things, the truth value of the belief being produced by the cognitive process and the fact that that is the belief being produced. Consequently, this indefinite probability must go the same way as objective definite probabilities and be either 1 or 0 depending upon whether the belief in question is true or false. Thus this reliabilist criterion entails the absurd consequence that in order for a belief to be justified it must be true.

Perhaps we can avoid this untoward result by appealing to reliability under less than totally specific circumstances. But how do we decide how specific to make the circumstances short of total specificity? The following seems initially hopeful. Although we cannot talk about the reliability of a cognitive process under the present circumstances in all their specificity without trivializing things, we can talk about the reliability under different general conditions satisfied by the present circumstances. Consider a belief P produced by the cognitive process M. If the present circumstances are of some general type C under which M is reliable, that might incline us to regard P as justified. For instance, M might be color vision and circumstances C might consist of viewing things in white light. But, given reliabilist intuitions, we would retract this judgment of justifiedness if the present circumstances were also of some more specific type C^* under which M is unreliable. For example, C^* might consist of viewing things in very dim white light. In other words, in evaluating M we need not appeal to totally specific circumstances, but we cannot ignore features of the present circumstances that would make M unreliable. This suggests that we should regard P as justified if and only if (1) there is a description C of the present circumstances such that M is reliable in circumstances of type C, and (2) there is no more specific description C^* of the present circumstances such that M is unreliable in circumstances of type C^*. Unfortunately, this does not resolve our problem. Suppose P is false. One condition satisfied by the present circumstances is that the process M is currently producing belief in the proposition P and P is false. The probability of a belief being true given that it is produced by M and M is currently producing belief in the proposition P and P is false is 0. Thus the reliabilist criterion once more entails that a belief is justified only if it is true.

   Rather than relativizing reliability to circumstances, reliabilists have tended to rely upon gerrymandering processes.[97] For instance, upon observing that color vision is unreliable in the universe at large, they might appeal to the gerrymandered process color-vision-in-white-light. It is important to realize that this is actually equivalent to relativizing reliability to circumstances and is subject to exactly similar problems. For example, for a reliabilist, justification would have to require not just that a belief be produced by a reliable process M, but also that there be no "more specific" process M^* (e.g., color-vision-in-dim-white-light) such that the belief is also produced by M^* and M^* is unreliable. But then we can reproduce the above argument, noting that if a belief P is true and it is produced by a process M then it is also produced by the gerrymandered process M-under-circumstances-in-which-the-belief-produced-is-P-and-P-is-true. Analogously, if P is false and it is produced by M then it is also produced by the gerrymandered process M-under-circumstances-in-which-the-belief-produced-is-P-and-P-is-false. Given any cognitive process producing P, there is a more specific gerrymandered process of one of these two sorts that also produces P, and it is reliable if and only

---

97. See Goldman's own discussion of what he calls the problem of generality in his (1979). Also see Richard Feldman (1986).

if *P* is true. So again we get the absurd result that a belief is justified if and only if it is true. The only way to avoid this is to impose some kind of restriction on gerrymandering, but there does not appear to be any non-ad hoc way of doing that.

We can see no way around this problem. In evaluating reliability relative to circumstances (or in gerrymandering processes) it cannot be reasonable to appeal to less specific descriptions of the circumstances in preference to more specific descriptions. That would amount to gratuitously throwing away information. But then the above problem is unavoidable. What is really happening here is that in trying to use reliability (which is an indefinite probability) to evaluate individual beliefs we are encountering the problem of direct inference all over again, and just as before[98], there is no way to obtain an objective assessment of the individual belief short of its truth value. Reliability is an indefinite probability and there is no way to get an objective definite probability out of it, but only an objective definite probability would be of any ultimate use to process reliabilism. It follows that process reliabilism is essentially bankrupt.

A final objection to process reliabilism concerns its motivation. Externalists were originally motivated by the intuition that in evaluating beliefs, what we want is to ensure that our beliefs are probable. Probabilist theories, proceeding as they do in terms of the definite probabilities of beliefs, attempt to capture this intuition fairly directly. Process reliabilism immediately diverges from the original intuition somewhat just by ceasing to talk about the probabilities of beliefs and talking instead about the indefinite probabilities of cognitive processes. However, reliabilists clearly wish to adhere to the original motivation of offering a deep and simple account of justification. Alvin Goldman (1979) writes,

> Since I seek an explanatory theory, i.e., one that clarifies the underlying source of justificational status, it is not enough for a theory to state "correct" necessary and sufficient conditions. Its conditions must also be appropriately deep or revelatory. Suppose, for example, that the following sufficient condition of justified belief is offered: 'If S senses redly at *t* and S believes at *t* that he is sensing redly, then S's belief at *t* that he is sensing redly is justified.' This is not the kind of principle I seek; for, even if it is correct, it leaves unexplained *why* a person who senses redly and believes that he does, believes this justifiably.

The problem is that once enough structure is added to process reliabilism to accommodate the difference between belief-dependent and belief-independent processes, defeasible reasoning, and so on, the resulting theory becomes at least as complicated as an internalist theory. If we can object to internalist theories that they are ad hoc attempts to produce criteria that pick out the intuitively right beliefs as justified, we can surely object to process reliabilism on the same grounds. Such reliabilist

---

98. See section 2.2, above.

theories provide no greater insight into what epistemic justification is all
about than do their internalist competitors.

4.3 Other Forms of Reliabilism

Externalist theories take the justifiability of a belief to be determined
in part by external considerations. Reliabilists propose that central among
those external considerations is some kind of reliability. Existing reliabilist
theories are all versions of process reliabilism, which appeals to the reliability of cognitive processes. The above methodological problems seem
to us to constitute a decisive refutation of process reliabilism. There is
simply no way to coherently formulate such a theory. This has gone
overlooked because philosophers have been content to make sloppy use
of probability, and that must be rectified.

We must not be too hasty in concluding that because process reliabilism
cannot be made to work, no form of reliabilism can be made to work.
No other forms of reliabilism have actually been proposed,[99] but that
does not mean that other forms of reliabilism are impossible. They
would appeal to the reliability of something other than cognitive processes.
At this point it may be hard to imagine what such an alternative form of
reliabilism might look like, but it will emerge in the next chapter that
there is a quite natural theory of this sort that will have to be taken
seriously.

5. Other Versions of Externalism

It is worth recalling at this point that externalism was also originally
recommended as a way of avoiding the ad hoc character of internalism.
But as externalist theories are made increasingly complex, they can no
longer be regarded as capturing the simple idea that in forming beliefs
what we should be trying to do is make it probable that our beliefs are
true. Externalist theories become just as ad hoc as their internalist competitors, although in a different way. The internalist theories are ad hoc
because they proceed piecemeal, propounding an array of unconnected
epistemic principles whose only recommendation is that they seem required to legitimize our intuitive reasoning. Complex externalist theories
avoid this difficulty by proposing a general analysis of epistemic justification from which all epistemic rules should be derivable. But the complex
analyses proposed by such theories are themselves ad hoc for much the
same reason as the internalist theories—the only motivation for the details
of the analyses is that they (hopefully) allow the analyses to avoid intuitive
counterexamples and capture what is intuitively the right reasoning. It
seems that despite initial appearances, any externalist theory that has

---

99. Goldman (1981) hints at a version of what we will call "norm reliabilism"
in the next chapter, but he does not formulate it clearly.

any hope of being correct is going to be just as ad hoc as the internalist theories it seeks to replace.

To some extent, the original motivation behind externalism has been recently forgotten or ignored and has been replaced by an alternative set of considerations. The externalist's attention to cognitive processes, fueled by the specific proposals of reliabilism, has invited epistemologists to consider the relevance of the abundant research on cognition that is carried out in the social and behavioral sciences. Since this research is situated in a framework where human beings are viewed as naturally evolved information processors, the involvement of the science of cognition has been seen as a way for externalists to advance a naturalized epistemology. Further, this has led to an ever-widening range of research that externalism-minded philosophers have tried to bring to bear in order to assess the formation of beliefs. This latter development typically follows in the wake of dissatisfaction with the insights generated by cognitive science.

It must be acknowledged that a correct externalist analysis of epistemic justification would be a significant achievement. But with the failure of any *simple* analyses there ceases to be any compelling reason to believe that a correct externalist analysis is possible. We believe that the remarks in the previous sections should be regarded as having successfully disposed of all existing externalist theories, but the criticisms presented are specific to the theories discussed and do nothing to refute externalism as such. Moreover, in light of the implicit reorientation in motivation behind some externalist theories, it remains an open question whether some other kind of externalist theory may be defensible. Although externalists have been some of the most visible proponents of naturalism, the motivation behind naturalized epistemology can be detached from reliabilist and externalist theories. In light of the considerable difficulties with externalism once it is detached from its original and intuitive motivation, the best hope for a naturalistic epistemology is to abandon externalism. This question will be taken up in the next two chapters, where it will eventually be concluded that no externalist theory can be correct and the true epistemological theory must be a naturalistic form of direct realism.

Finally, we should mention a theory that is externalist but seems motivated by neither the intuitiveness of probabilism nor the drive to naturalism. Alvin Plantinga has defended a novel view that he calls the theory of proper functions (1988 and 1993b). Plantinga's theory was motivated by his desire to solve problems with other theories that arise when due attention is paid to the possibility of cognitive malfunctions. For instance, he objects to process reliabilism because it allows for a cognitive process to be reliable as a result of a failure to function properly. Plantinga alleges that such reliability intuitively does not confer positive epistemic status. Imagine that I have a brain tumor whose only effect is to cause me to believe that I have a brain tumor. Suppose further that I have every reason to believe that I do not have a tumor because I have been assured of my complete health by experts. This belief is the product of a highly reliable process and there is no other more reliable process that would result in my believing that I do not have a brain tumor. Any more reliable process would reveal that I do, indeed, have a brain tumor. Still, my belief in my brain tumor does not seem to have positive epistemic status.

This is just another allegation to the effect that reliabilism is unintuitive. Plantinga's innovation is to diagnose that unintuitiveness and to propose a positive view on the basis of that diagnosis. In his theory, Plantinga claims that, in order to confer positive epistemic status,[100] a cognitive process must be working properly in an environment for which it is appropriate. These two clauses make reference to a design plan that specifies what "proper function" means for human cognitive processes and specifies the appropriate environment for cognitive processes.[101] The design plan also enumerates the cognitive processes that are aimed at producing true beliefs, as these are the only ones that can confer positive epistemic status.

There are numerous difficulties associated with Plantinga's proposals having to do with proper function and design plans. He is aware of many of them and has heroically attempted to offer responses, especially in his (1993b). We will simply note here that all of the information about our actual design plan and proper functioning (should there be such things) are external to the epistemic agent. The theory of proper functions will be dealt with, then, if it can be shown that no version of externalism is true. That will be attempted in chapter five.

---

100. We think that "positive epistemic status" is Plantinga's phrase for justification in his (1988) (see pp. 2-3). This article is the early formulation of Plantinga's view. His official reason for the change in terminology is that "justification" suggests a deontological conception of epistemic achievement, which Plantinga thinks is up for grabs. In the later formulations his theory (1993a) and (1993b), "warrant" replaces positive epistemic status, and appears to have more to do with whatever, in addition to belief and truth, yeilds knowledge. On that construal of warrant, we are not as directly interested in the theory of proper functions. On the other hand, some of Plantinga's remarks are sufficiently ambiguous to suggest that the theory of proper function might be used as a candidate theory of justification. Our remarks are meant to apply to this possibility, even if it is not the one envisioned by Plantinga.

101. Plantinga takes the design plan to be of divine origin.

5
EPISTEMIC NORMS

1. Recapitulation

We have surveyed existing theories of knowledge and concluded that most are subject to fatal objections. Doxastic theories, both foundationalist and coherence, fail because they cannot accommodate perception and memory. These are cognitive processes that produce beliefs in us, and the beliefs are sometimes justified and sometimes unjustified, but whether they are justified is not just a function of one's other beliefs. It follows that justifiability is a function of more than doxastic states, and hence the true epistemological theory must be a nondoxastic theory. Nondoxastic theories can be internalist or externalist. We have sketched an internalist nondoxastic theory—direct realism—and one of our ultimate purposes in this book is to defend a variety of direct realism. This will be done by arguing against externalist theories. If all externalist theories can be rejected, the only remaining theories are internalist nondoxastic theories, and we take it that direct realism is the most plausible such theory. The premiere externalist theories are all versions of either probabilism or process reliabilism. These theories fail for a variety of reasons specific to them. But it remains possible at this stage that some other form of externalism might succeed, so a more general argument against externalism is required if we are to defend direct realism in this way.

On another front, all of the theories thus far discussed are subject to a common objection. This is that they fail to give illuminating general accounts of epistemic justification. Although they may start with simple and intuitive ideas, when confronted with detailed objections they are forced to complicate those simple ideas and, in the end, they propound complex and convoluted criteria of justifiedness. Even if some such complex criterion were correct, we would be left wondering why such a concept of epistemic justification should be of interest to us. It has been objected that foundations theories and direct realism propose ad hoc lists of epistemic rules whose only defense is that they seem to be required for the justifiedness of those beliefs we antecedently regard as justified. As formulated, those theories give no principled account of epistemic justification from which this medley of rules might be derived. But we have found that much the same objection can be raised to all the other theories we have discussed as well. The final versions of these theories leave us with such complicated criteria that they cannot be regarded as explanations of what epistemic justification is all about.

To sort this out we need a general account of epistemic justification, and it will be the purpose of this chapter to provide such an account.

Once we have a better understanding of epistemic justification it will become possible to dismiss all externalist theories for deep reasons having to do with the general nature of epistemic justification. Basically the same considerations will also necessitate the rejection of a wide variety of internalist theories, including most coherence theories. Ultimately, we aim to offer a simple and explanatory theory of what justification is. The general account of epistemic justification that will be proposed here and in the next chapter has the further virtue that it is a naturalistic account, in the sense that it integrates the concept of epistemic justification into a naturalistic view of human beings as biological machines.

2. Epistemic Norms

What are we asking when we ask whether a belief is justified? What we want to know is whether it is all right to believe it. Justification is a matter of "epistemic permissibility". It is this normative character of epistemic justification that we want to emphasize. That epistemic justification is a normative notion is not a novel observation. The language of epistemic justification is explicitly normative, and a recurrent theme has been that justification is connected with the "ethics of belief". This has played a role in the thought of a number of epistemologists. Roderick Chisholm (1977 and chapter one of 1957) has repeatedly stressed the normative character of epistemic terms, several recent philosophers have proposed analyzing epistemic justification in terms of the maximization of epistemic values,[102] and a few philosophers have appealed to the normative character of justification in other ways.[103] Thus we will think of epistemic justification as being concerned with questions of the form, "When is it permissible (from an epistemological point of view) to believe P?" This is the concept of epistemic justification that we are concerned to explore.

Norms are general descriptions of the circumstances under which various kinds of normative judgments are correct. Epistemic norms are norms describing when it is epistemically permissible to hold various beliefs. A belief is justified if and only if it is licensed by correct epistemic norms. We assess the justifiedness of a belief in terms of the cognizer's reasons for holding it, and our most fundamental epistemic judgments pertain to reasoning (construing reasoning in the broad manner required by direct realism). Thus we can regard epistemic norms as the norms governing "right reasoning". Epistemic norms are supposed to guide us in reasoning and thereby in forming beliefs. The concept of epistemic

---

102. See for example Isaac Levi (1967), Keith Lehrer (1974, p. 146ff and 204ff) and (1981, p. 75ff), and Alvin Goldman (1981, pp. 27-52).
103. See for example Hilary Kornblith (1983). See also William Alston (1978), Roderick Firth (1978), John Heil (1983), and Jack Meiland (1980).

justification can therefore be explained by explaining the nature and origin of the epistemic norms that govern our reasoning. We have been calling this "the procedural concept of epistemic justification". There may be other concepts that can reasonably be labeled "epistemic justification", but it is the procedural concept that is the focus of the present book and is involved in traditional epistemological problems.

Much of recent epistemology has been concerned with describing the contents of our epistemic norms, but the nature and source of epistemic norms has not received much attention. Epistemologists have commonly supposed that epistemic norms are much like moral norms and that they are used in evaluating reasoning in the same way moral norms are used in evaluating actions. One of the main contentions of this chapter will be that this parallel is not at all exact and that epistemologists have been misled in important ways by supposing the analogy to be better than it is.[104] A proper understanding of epistemic norms will provide us with a radically new perspective on epistemology, and from the point of view of this perspective new light can be shed on a number of central epistemological problems.

3. How Do Epistemic Norms Regulate?

3.1 Epistemic Normativity

In order to get a grasp of the nature of epistemic norms, let us begin by asking their purpose. It is important to distinguish between two uses of norms (epistemic or otherwise). On the one hand, there are third-person uses of norms wherein we use the norms to evaluate the behavior of others. Various norms may be appropriate for third-person evaluations, depending upon the purpose we have in making the evaluations. For example, we may want to determine whether a person is a good scientist because we are trying to decide whether to hire her. To be contrasted with third-person uses of norms are first-person uses. First-person uses of norms are, roughly speaking, action-guiding.[105] For example, I might appeal to *Fowler's Modern English Usage* to decide whether to use 'that' or 'which' in a sentence. We will call such action-guiding norms "procedural". Epistemological questions are about rational cognition—about how cognition rationally ought to work—and so are inherently first-person. The traditional epistemologist asks, "How is it possible for me to be justified in my beliefs about the external world, about other minds, about the past, and so on?" These are questions about what to believe. Epistemic norms are the norms in terms of which these questions are to be answered, so these norms are used in a first-person reason-guiding or procedural capacity.

---

104. See Michael DePaul (1993) for some discussion of this.
105. We can also make "third-person evaluations" of our own past behavior, but that is different from what we are calling "first-person uses" of norms.

3.2 The Intellectualist Model

If reasoning is governed by epistemic norms, just how is it governed? There is a model of this regulative process that is often implicit in epistemological thinking, but when we make the model explicit it is *obviously* wrong. This model assimilates the functioning of epistemic norms to the functioning of explicitly articulated norms. For example, naval officers are supposed to "do it by the book", which means that whenever they are in doubt about what to do in a particular situation they are supposed to consult explicit regulations governing all aspects of their behavior and act accordingly. Explicitly articulated norms are also found in driving manuals, etiquette books, and so on. Without giving the matter much thought, there is a tendency to suppose that all norms work this way, and in particular to suppose that this is the way epistemic norms work. We will call this "the intellectualist model".[106] It takes little reflection to realize that epistemic norms cannot function in accordance with the intellectualist model. If we had to make an explicit appeal to epistemic norms in order to acquire justified beliefs we would find ourselves in an infinite regress, because to apply explicitly formulated norms we must first acquire justified beliefs about how they apply to this particular case. For example, if we are to reason by making explicit appeal to a norm telling us that it is permissible to move from the belief that something looks red to us to the belief that it is red, we would first have to become justified in believing that that norm is included among our epistemic norms and we would have to become justified in believing that we believe that the object looks red to us. In order to become justified in holding those beliefs, we would have to apply other epistemic norms, and so on *ad infinitum*. Thus it is clear that epistemic norms cannot guide our reasoning in this way.[107]

3.3 Do Epistemic Norms Regulate?

If the intellectualist model is wrong, then how do epistemic norms govern reasoning? At this point we might raise the possibility that they do not. Perhaps epistemic norms are only of use in third-person evaluations. But it cannot really be true that epistemic norms play *no role at all* in first-person deliberations. We can certainly subject our reasoning to self-criticism. Every philosopher has detected invalid arguments in his or her own reasoning. This might suggest that epistemic norms are

---

106. Many philosophers appear to adopt the intellectualist model, although it is doubtful that any of them would seriously defend it if challenged. For example, Alvin Goldman (1981) appears to assume such an account of epistemic norms. The intellectualist model pervades Hilary Kornblith's (1983) discussion. Unfortunately, it is also prominent in Pollock's (1979) discussion.

107. This point has been made several times. Pollock made it in his (1974), and James van Cleve (1979) made it again. Despite this, we do not think that epistemologists have generally appreciated its significance.

only relevant in a negative way. Our reasoning is innocent until proven guilty. We can use reasoning to criticize reasoning, and hence we can use reasoning in applying epistemic norms to other reasoning, but we cannot be required to reason about norms *before* we can do any reasoning. This would avoid the infinite regress.

As theoretically attractive as the "innocent until proven guilty" picture might be, it cannot be right. It entails the view, already discussed and rejected in chapter three, according to which all beliefs are prima facie justified. This view cannot handle the fact that epistemic norms guide the acquisition of beliefs and not just their after-the-fact evaluation. This was illustrated by the observation that even in the perceptual acquisition of beliefs about physical objects, the resulting beliefs are sometimes unjustified. More generally, there are a number of natural processes that lead to belief formation. Among these are such "approved" processes as vision, inductive reasoning, deductive reasoning, and memory, and also some "unapproved" but equally natural processes such as wishful thinking. The latter is just as natural as the former. Recall the example given earlier. My daughter had gone to a football game, the evening had turned cold, and I was worried about whether she took a coat. I found myself thinking, "Oh, I am sure she is wearing a coat". But then on reflection I decided that I had no reason to believe that—my initial belief was just a matter of wishful thinking. The point here is that wishful thinking is a natural belief-forming process, but we do not accord it the same status as some other belief-forming processes like vision. Although we have a natural tendency to form beliefs by wishful thinking, we also seem to "naturally" know better. This is not just a matter of after-the-fact criticism. We know better than to indulge in wishful thinking at the very time we do it. It seems that *while* we are reasoning we are being guided by epistemic norms that preclude wishful thinking but permit belief formation based upon perception, induction, and so on. This is of more than casual significance, because it might be impossible to rule out wishful thinking by after-the-fact reasoning. This is because the after-the-fact reasoning might include wishful thinking again, and the new wishful thinking could legitimize the earlier wishful thinking. If epistemic norms play no regulative role in our reasoning while it is going on, there is no reason to think they will be able to play a successful corrective role in after-the-fact evaluations of reasoning. In order for the corrective reasoning to be successful it must itself be normatively correct. Epistemic norms must, and apparently do, play a role in guiding our epistemic behavior at the very time it is occurring. But how can they?

Epistemic norms cannot play a merely negative, corrective, role in guiding reasoning, nor can they function in a way that requires us to already make judgments before we can make judgments. What is left? Our perplexity reflects an inadequate understanding of the way procedural norms usually function. The case of making an explicit appeal to norms in order to decide what to do is the exception rather than the rule. You may make reference to a driving manual when you are first learning to drive a car, but once you learn how to drive a car you do not look things up in the manual anymore. You do not usually give any explicit thought to what to do—you just do it. This does not mean, however, that your behavior is no longer guided by those norms you learned when you first learned to drive. Similarly, when you first learned to ride a bicycle you were told to turn the handlebars to the right when the bicycle leaned to the right. You learned to ride in accordance with that norm, and that norm still governs your bike riding behavior but you no longer have to think about it. The point here is that norms can govern your behavior without your having to think about them. The intellectualist model of the way norms guide behavior is almost always wrong. This point has been insufficiently appreciated. It is of major importance in understanding epistemic norms. Reasoning is more like riding a bicycle than it is like being in the navy.

3.4 Procedural Knowledge

What makes it possible for your bike-riding behavior to be governed by norms without your thinking about the norms is that you *know how* to ride a bicycle. This is *procedural knowledge* rather than *declarative knowledge*. Having procedural knowledge of what to do under various circumstances does not involve being able to give a general description of what we should do under those circumstances. This is the familiar observation that knowing how to ride a bicycle does not automatically enable one to write a treatise on bicycle riding. This is true for two different reasons. First, knowing how to ride a bicycle requires us to know what to do in each situation *as it arises*, but it does not require us to be able to say what we should do before the fact. Second, even when a situation has actually arisen, our knowing what to do in that situation need not be propositional knowledge. In the case of knowing that we should turn the handlebars to the right when the bicycle leans to the right, it is plausible to suppose that most bicycle riders do have propositional knowledge of this; but consider knowing how to hit a tennis ball with a tennis racket. We know how to do it—as the situation unfolds, at each instant we know what to do—but even at that instant we cannot give a description of what we should do. Knowing what to do is the same thing as knowing to do it, and that need not involve propositional knowledge.

We can give a rough description of how procedural norms govern behavior in a non-intellectualist manner. When we learn how to do something X, we "acquire" a plan of how to do it. That plan might (but need not) start out as explicit propositional knowledge of what to do under various circumstances, but then the plan becomes internalized. Using a computer metaphor, psychologists sometimes talk about procedural knowledge being "compiled-in". When we subsequently undertake to do X, our behavior is automatically channeled into that plan. This is just a fact of psychology. We form habits or conditioned reflexes. Norms for doing X constitute a description of this plan for doing X. The sense in which the norms guide our behavior in doing X is that the norms describe the way in which, once we have learned how to do X, our behavior is automatically channeled in undertaking to do X. The norms are not, however, just descriptions of what we do. Rather, they are descriptions of what we *try* to do. Norms can be hard to follow and we follow them with varying degrees of success. Think, for example, of an expert golfer who knows how to swing a golf club. Nevertheless, he does not always get his stroke right. It is noteworthy, and it will be important later, that when he does not get his stroke right he is often able to tell that by something akin to introspection. When he does it wrong it "feels wrong". The ability to tell in this way whether one is doing something right is particularly important for those skills governing performances (like golf swings) that take place over more than just an instant of time, because it enables us to correct or fine tune our performance as we go along.

The internalization of norms results in our having "automatic" procedural knowledge that enables us to do something without having to think about how to do it. It is this process that we are calling "being guided by the norm without having to think about the norm". This may be a slightly misleading way of talking, because it suggests that somewhere in our heads there is a mental representation of the norm and that mental representation is doing the guiding. Perhaps it would be less misleading to say that our behavior is being guided by our procedural knowledge and the way in which it is being guided is described by the norm. What is important is that this is a particular way of being guided. It involves non-intellectual psychological mechanisms that both guide and correct (or fine tune) our behavior.

3.4.1 The Competence/Performance Distinction

The distinction between knowing how to do something and actually doing it is the same as the competence/performance distinction in linguistics. When linguists study a language, they try to discover what the rules are that determine which utterances are grammatical. But it is a contingent matter what the rules are that govern any given language, and those rules may change over time as the language evolves. The language is determined by the way the speakers of the language use it. However, eliciting the rules of grammaticality for a language is not the same thing as simply describing how the users of the language talk. Linguists observe that many, perhaps most, of our utterances are ungrammatical. Our speech is populated with "Ahh"s and "Umm"s, we leave sentences unfinished, and commit a variety of other grammatical infractions. But we know better. When linguists study language, they are not interested in what we do when we do it wrong. Linguistic theories are about what we do when we do it right. However, right and wrong in this case are not determined by some metaphysically necessary standard. They are determined by how people talk. To prevent the account from going around in a circle, Noam Chomsky (1965) proposed that what people know when they know how to speak a language takes the form of procedural knowledge. They *know how* to speak the language. As in the case of any procedural knowledge, people can have internalized rules for how to speak the language but violate them. The objective of the linguist is to describe the internalized rules that comprise the speakers' procedural knowledge. Chomsky referred to this as a *competence theory* of language, and contrasted it with a *performance theory*, which would be a theory of what people actually say, errors and all. A performance theory might be the kind of theory that a psychologist would produce, but linguists seek a competence theory.

3.4.2 Normative Language

The use of normative language in formulating procedural norms is pervasive. Norms can be described as knowing what we *should* do under particular circumstances. The point of using normative language to describe internalized norms is to contrast what the norms tell us to do with what we *do*. The simple fact of the matter is that even when we know how to do something (e.g., swing a golf club) we do not always succeed in following our norms. This use of 'should' in describing procedural knowledge is interesting. Moral philosophers have talked about different senses of 'should', distinguishing particularly between moral uses of 'should' and goal-directed uses of 'should'. An example of the latter is "If you want the knife to be sharp then you should sharpen it on the whetstone". But the use of 'should' in "In riding a bicycle, when the bicycle leans to the right you should turn the handlebars to the right" is of neither of these varieties. It is perhaps more like the goal-directed kind of 'should', but we are not saying that that is what you should do to achieve the goal of riding a bicycle. Rather, that is part of what is involved *in* riding a bicycle—that is *how* to ride a bicycle.

Note that a similar use of normative language occurs in formulating rules of grammar. We are informed that under certain circumstances we should say 'that' rather than 'which'. The 'should' here is not a moral or prudential should. It is that kind of 'should' that we use in describing procedural knowledge. Because it is natural to use normative language in describing procedural knowledge, it is equally natural to say that in acquiring procedural knowledge, what we learn are norms for how we should do something.

3.4.3 Epistemic Norms Are Procedural Norms

So far we have been describing procedural norms in general. Now let us apply these insights to epistemic norms. We know how to reason, or more generally, how to cognize. That means that under various circumstances we know what to do in cognizing. This can be described equivalently by saying that we know what we should do. Our epistemic norms are just the norms that describe this procedural knowledge, and rational cognition is cognition in compliance with the norms. The way epistemic norms can guide our cognition without our having to think about them is no longer mysterious. They describe an internalized pattern of behavior that we automatically follow in epistemic cognition, in the same way we automatically follow a pattern in bicycle riding. Epistemic norms are the internalized norms that govern our epistemic cognition. Once we realize that they are just one more manifestation of the general phenomenon of automatic behavior governed by internalized norms, epistemic norms should no longer seem puzzling. The mystery surrounding epistemic norms evaporates once we recognize that the governing process is a general one and its application to epistemic norms and epistemic cognition is not much different from its application to any other kind of procedural norms. Of course, unlike most norms our epistemic norms may be innate, in which case there is no process of internalization that is required to make them available for use in guiding our cognition.

There has been a great deal of recent work in psychology concerning human irrationality. Psychologists have shown that in certain kinds of epistemic situations people have an almost overpowering tendency to reason incorrectly.[108] It might be tempting to conclude from this that, contrary to what we are claiming, people do not know how to reason.[109] The short way with this charge is to note that if we did not know how to reason correctly in these cases, we would be unable to discover that people reason incorrectly. To say that we know how to reason is to invoke a competence/performance distinction. It is no way precludes our making mistakes. It does not even preclude our almost always making mistakes in specific kinds of reasoning. All it requires is that we can, in principle, discover the errors of our ways and correct them.[110]

4. The Refutation of Externalism

We have described how our epistemic norms work. But this is not yet to say anything about which epistemic norms are correct. An episte-mological theory must answer two different questions. First, it must describe the correct epistemic norms. Second, it must tell us what makes them correct. The first question concerns the content of epistemic norms,

---

108. Much of the central psychological material can be found in Daniel Kahneman, Paul Slovic, and Amos Tversky (1982), and R. E. Nisbett and L. Ross (1980). For an overview, see Massimo Piatelli-Palmarini (1994).

109. Murray Clarke (1990) criticizes our view on these grounds.

110. This is pretty much the same as the assessment of the irrationality literature offered by Jonathan Cohen (1981). See also the critique in Alvin Goldman (1986). Edward Stein (1995) offers a thorough and illuminating discussion of the philosophical import of psychological research on reasoning. It should be noted here that not all psychological research is pessimistic about the performance of human reasoning. See Gerd Gigerenzer (1991) and (1996).

and the second question concerns their justification. By distinguishing between these questions we can see the internalism/externalism distinction in a new light. A belief is justified if and only if it is held in compliance with correct epistemic norms. Externalism is the view that the justifiedness of a belief is a function in part of external considerations. Thus if externalism is right, external considerations must play a role in determining whether a belief is held in compliance with correct epistemic norms. This could arise in either of two ways. On the one hand, external considerations could enter into the formulation of correct epistemic norms. On the other hand, it might be granted that epistemic norms can only appeal to internal considerations, but it might be insisted that external considerations are relevant to determining which set of internalist norms is correct. Thus we are led to a distinction between two kinds of externalism. *Belief externalism* insists that correct epistemic norms must be formulated in terms of external considerations. A typical example of such a proposed norm might be "It is permissible to hold a belief if it is generated by a reliable cognitive process." In contrast to this, *norm externalism* acknowledges that the content of our epistemic norms must be internalist, but employs external considerations in the selection of the norms themselves. The distinction between belief and norm externalism is analogous to the distinction between act and rule utilitarianism. Externalism (simpliciter) is the disjunction of belief externalism and norm externalism. In the last chapter, we were concerned with dealing with proposals of the first sort. At the end of that discussion, we noted that there was another sense of externalism that we would eventually have to grapple with, namely norm externalism. A number of philosophers who are normally considered externalists appear to vacillate between belief externalism and norm externalism.[111] The difference between these two varieties of externalism will prove important. In the end, both must be rejected, but they are subject to different difficulties.[112]

According to internalism, the justifiedness of a belief is a function exclusively of internal considerations, so internalism implies the denial of both belief and norm externalism. That is, the internalist maintains that epistemic norms must be formulated in terms of relations between beliefs or between beliefs and nondoxastic internal states (e.g., perceptual states), and she denies that these norms are subject to evaluation in terms of external considerations. Typically, the internalist has held that

---

111. Alvin Plantinga (1993a) is guilty of this. He criticizes the view that we are defending in this book by assuming that some version of norm externalism can be made to work (see especially pp. 171-6). He fails to recognize that norm externalism would have to be defended separately from his endorsement of the Theory of Proper Functions, which is a kind of belief externalism.

112. Alvin Goldman (1981) seems to be one of the few externalists who is clear on this distinction. He distinguishes between two senses of 'epistemic justification' and adopts belief externalism with regard to one and norm externalism with regard to the other.

whatever our *actual* epistemic norms are, they are necessarily correct and not subject to criticism on any grounds (externalist or otherwise). Of course, this is precisely where internalists disagree with norm externalists. Let us turn then to a reconsideration of both forms of externalism in the light of our new understanding of epistemic norms.

4.1 Belief Externalism

Now that we understand how epistemic norms work in guiding our epistemic cognition, it is easy to see that they must be internalist norms. This is because when we learn how to do something we acquire a set of norms for doing it and these norms are internalized in a way enabling our cognitive system to follow them in an automatic way without our having to think about them. This has implications for the content of our norms. For example, we have been describing one of our bike-riding norms as telling us that if the bicycle leans to the right then we should turn the handlebars to the right, but that is not really what we learn when we learn to ride a bicycle. The automatic processing systems implemented in our neurology do not have access to whether the bicycle is leaning to the right. What they do have access to are things like our *thinking* that the bicycle is leaning to the right, and certain balance sensations emanating from our inner ear. What we learn is (roughly) to turn the handlebars to the right if we either experience those balance sensations or think on some other basis that the bicycle is leaning to the right. The circumstance-types to which our norms appeal in telling us to do something in circumstances of those types must be directly accessible to our system of cognitive processing. The sense in which they must be directly accessible is that our cognitive system must be able to access them without our first having to make a *judgment* about whether we are in circumstances of that type. We must have non-epistemic access.[113]

This general observation about procedural norms has immediate implications for the nature of our epistemic norms. It implies, for example, that epistemic norms cannot appeal to external considerations of reliability. This is because such norms could not be internalized. Like *the bicycle's leaning to the right*, considerations of reliability are not directly accessible to our automatic processing systems. There is in principle no way that we can learn to make inferences of various kinds only if they are *in fact* reliable. Of course, we could learn to make certain inferences only if we *think* they are reliable, but that would be an internalist norm appealing

---

113. It might be insisted that this is at least sometimes a misleading way of talking—if our norms for doing X tell us to do Y whenever we *think* it is the case that C, we might better describe our norms as telling us to do Y when it *is* the case that C. We do not care if one chooses to talk that way, but it must be realized that it has the consequence that although the reformulated norm says to do Y when it is the case that C, knowing how to do X will really only result in our doing Y when we *think* it is the case that C. This will be important. (And, of course, norms appealing to internal states other than beliefs could not be reformulated in this manner anyway.)

to *thoughts* about reliability rather than an externalist norm appealing to reliability itself.[114] Similar observations apply to any externalist norms. Consequently, it is in principle impossible for us to employ externalist norms. We take this to be a conclusive refutation of belief externalism.

We introduced the internalism/externalism distinction by saying that internalist theories make justifiedness a function exclusively of the believer's internal states, where internal states are those that are "directly accessible" to the believer. The notion of direct accessibility was purposely left vague, but it can now be clarified. We propose to define internal states as those states that are directly accessible to the cognitive mechanisms that direct our epistemic cognition. The sense in which they are *directly* accessible is that access to them does not require us first to have beliefs about them. This definition makes the internalist/externalist distinction precise in a way that agrees at least approximately with the way it has generally been used, although it is impossible to make it agree with everything everyone has said about it because philosophers have drawn the distinction in different ways. It especially noteworthy, however, that our access constraint is considerably more liberal than the reflective access required by some internalist theories. Reflective access seems too restrictive in light of the way that norms automatically govern cognition. The internalism/externalism distinction will be discussed further when we reflect on the status of naturalism in epistemology.

We have characterized internalist theories in terms of direct accessibility, but we have not said anything in a general way about which properties and relations are directly accessible. It seems clear that directly accessible properties must be in some sense "psychological", but it is doubtful that we can say much more than that from the comfort of our armchairs. That is an empirical question to be answered by psychologists. Despite the fact that we do not have a general characterization of direct accessibility, it is perfectly clear in many specific cases that particular properties to which philosophers have appealed are not directly accessible. In light of this, the preceding refutation of belief externalism can be applied to a remarkably broad spectrum of theories, and it seems to us to constitute an absolutely conclusive refutation of those theories. We have indicated how it applies to theories formulating epistemic norms in terms of reliability. It applies in the same way to probabilist theories. For example, we saw that many probabilists endorse the *simple rule*:

A belief is epistemically permissible if and only if what is believed is sufficiently probable.

---

114. It would also be a wholly implausible theory. We do not invariably have beliefs about the reliability of our inferences whenever we make them, and if norms *requiring* us to have such beliefs also require those beliefs to be justified then they lead to an infinite regress.

If the simple rule is to provide us with a procedural norm then the probability of a belief must be a directly accessible property of it. No objective probability can have that property. Thus it is impossible to use the simple rule, interpreted in terms of objective probabilities, as a procedural norm. This objection could be circumvented by replacing the simple rule with its "doxastic counterpart":

A belief is epistemically permissible if and only if the epistemic agent believes it to be highly probable.

But this rule formulates an internalist norm (albeit, an implausible one[115]). It might be supposed that we could breath life back into the simple rule by interpreting it in terms of subjective probability. Here we must be careful to distinguish between subjective probability as actual degree of belief and subjective probability as rational degree of belief. Interpreted in terms of actual degrees of belief, the simple rule would amount to the claim that a belief is justified if and only if it is firmly held, which is an internalist norm, but a preposterous one. To get a plausible norm, we must interpret subjective probability as rational degrees of belief. Rational degree of belief is the unique degree of belief one rationally ought to have in a proposition given one's overall doxastic state, and this is to be understood in terms of prudentially rational betting behavior. As we have indicated, we have serious doubts about the intelligibility of this notion. But even if we waive that objection, ascertaining what this unique rational degree of belief should be is immensely difficult. It seems extremely unlikely that the rational degree of belief one ought to have in a proposition is a directly accessible property of it. If it is not then this version of the simple rule also succumbs to our general objection to belief externalism.

Other epistemological theories succumb to this objection to belief externalism. For example, Keith Lehrer's coherence theory is an internalist theory, but it was pointed out in the last chapter that an externalist theory can be modeled on it. According to this externalist theory, a person is justified in believing a proposition if and only if that proposition is more probable than each proposition competing with it. But a proposition's being more probable than any of its competitors is not a directly accessible property of it, and hence the objective version of Lehrer's theory becomes incapable of supplying us with a procedural norm.

The net cast by this objection catches some internalist theories as well. For instance, a holistic coherence theory adopts a holistic view of reasons according to which a belief is licensed if it is suitably related to the set of all the beliefs one holds. A holistic coherence theory requires a

---

115. We do not ordinarily have any beliefs at all about the probabilities of what we believe. Furthermore, even if we did they would presumably not render our beliefs justified unless the probability beliefs were themselves justified, so we would be threatened by an infinite regress.

relationship between a justified belief and the set of all the beliefs one holds, but that will not normally be a directly accessible property of the justified belief, and hence although the norm proposed by the holistic theory will be an internalist norm, it will not be internalizable. Thus it cannot serve as a procedural norm.

The general point emerging from all this is that the norms proposed by many traditional theories cannot be reason-guiding. Accordingly, they cannot serve as epistemic norms. No non-internalist theory can provide us with epistemic norms that we could actually use. Correct epistemic norms must be internalist. On the other hand, we have also seen that epistemic norms must be able to appeal to more than the cognizer's doxastic state. They must also be able to appeal to his perceptual and memory states. Thus the correct epistemological theory must endorse some kind of nondoxastic internalist norm.

Is there any way to salvage belief externalism in the face of the objection that it cannot give reasonable accounts of first-person reason-guiding epistemic norms? The possibility remains that belief externalism might provide norms for third-person evaluations. We think it is noteworthy in this connection that externalists tend to take a third-person point of view in discussing epistemology. If externalist norms played a role in third-person evaluations, we would then have both externalist and internalist norms that could be applied to individual beliefs and they might conflict. What would this show? It would not show anything—they would just be different norms evaluating the same belief from different points of view. We can imagine a persistent externalist insisting, "Well, if the two sets of norms conflict, which way should we reason—which set of norms should we follow?" But that question does not make any sense. Asking what we should do is asking for a normative judgment, and before we can answer the question we must inquire to what norms the 'should' is appealing. To make this clearer consider an analogous case. We can evaluate beliefs from both an epistemic point of view and a prudential point of view. Recall Helen who has good reason for believing that her father is Jack the Ripper. Suppose that if she believed that, it would be psychologically crushing. Then we might say that, epistemically, she should believe it, but prudentially she should not. If one then insists upon asking, "Well, should she believe it or not?", the proper response is, "In what sense of 'should'—epistemic or prudential?" Similarly, if externalist and internalist norms conflict and one asks, "Which way should we reason?", the proper response is to ask to which set of norms the 'should' is appealing. The point is that different norms serve different purposes, and when they conflict that does not show that there is something wrong with one of the sets of norms—it just shows that the different norms are doing different jobs. The job of internalist norms is reason-guiding, and as such they are the norms traditionally sought in epistemology. Externalist norms (if any sense can be made of them) may also have a point, but they cannot be used to solve traditional epistemological problems pertaining to epistemic justification.

4.2 Reconsidering the Doxastic Assumption

The endorsement of nondoxastic norms amounts to the rejection of the doxastic assumption, but that has often seemed puzzling. How is it be possible for nondoxastic states to justify beliefs when we are not aware that we are in them? Recall the argument given for the doxastic assumption in chapter one. Procedural epistemic justification is supposed to be concerned with what to believe. But in deciding what to believe, we can only take account of something insofar as we have a belief about it. Thus only beliefs can be relevant to what we are justified in believing. We are now in a position to see what is wrong with that argument, and accordingly to understand how nondoxastic norms are possible. First notice that this argument for the doxastic assumption could not possibly be right, because it is self-defeating. If this argument were right, we could only take account of our beliefs insofar as we have beliefs about our beliefs, and then an infinite regress would loom. There has to be something about beliefs that makes them the sort of thing we can take account of without having beliefs about them. What could this be?

What is it to take account of something in the course of cognition? It is to use it in our cognitive deliberations. We can take account of anything by having a belief about it, but cognition has to start somewhere, with things that we don't have beliefs about. Obviously, it can start with beliefs. The reason it can start with beliefs is that they are internal states, and cognition is an internal process that can access internal states directly. Cognition works by noting that we have certain beliefs and using that to trigger the formation of further beliefs. However, it is *cognition* that must note that we have certain beliefs—*we* do not have to note it ourselves. The sense in which cognition notes it is metaphorical—it is the same as the sense in which a computer program accessing a database might be described as noting that some particular item is contained in it.

We have seen that epistemologists have a lamentable tendency to over-intellectualize cognition. Human beings are *cognitive machines*. We are unusual machines in that our machinery can turn upon itself and enable us to direct many of our own internal operations. Many of these operations, like reasoning, can proceed mechanically, without any deliberate direction or intervention from us, but when we take a mind to we can directly affect their course. For example, we can, at least to some extent, decide what to think about, decide not to pursue a certain line of investigation, and to pursue another one instead. There must, however, be a limit to the extent to which we are *required* to do this. After all, the processes by which we do it are a subspecies of the very processes in which we are intervening. If we had to explicitly direct all of our cognitive processes, we would also have to direct the ones involved in doing the directing, and we would again have an infinite regress.

The significance of this is that we don't *have* to think about our reasoning in order to reason. It is important, for various reasons, that we *can* think about it when the need arises, but we don't have to and don't usually do it. Thus reasoning can proceed by moving from beliefs to beliefs without our thinking about either reasoning or the beliefs. By virtue of doing the reasoning we are thinking about whatever the beliefs are about, not about the beliefs themselves. This explains the sense in which cognition can take account of our having certain beliefs without our having to have beliefs to the effect that we have those beliefs. But note that it explains much more. In precisely the same sense, cognition can take account of other internal states, for example, percepts, without our having to have beliefs to the effect that we are in those states. Thus there is no reason why cognition cannot move directly from percepts to beliefs about the physical objects putatively represented by the percepts.

Cognition can make use of any states to which it has direct access, but those are just the internal states. So cognition can make use of any internal states without our having beliefs about those states, and correspondingly our epistemic norms can appeal to any internal states—not just beliefs. Such nondoxastic norms only seemed puzzling because we were implicitly assuming the intellectualist model of the way epistemic norms regulate belief. Given the way epistemic norms actually operate, all that is required is that the input states be directly accessible. Belief states are directly accessible, but so are a variety of nondoxastic states like perceptual states and memory states. Thus there is no reason why epistemic norms cannot appeal to those states, and the rejection of the doxastic assumption and the move to direct realism ceases to be puzzling.

4.3 Norm Externalism

Recall that there are two kinds of externalism. Belief externalism advocates the adoption of externalist norms. We regard belief externalism as having been decisively refuted by the preceding considerations. Norm externalism, on the other hand, acknowledges that we must employ internalist norms in our reasoning, but proposes that alternative sets of internalist norms should be evaluated in terms of external considerations. For example, it may be alleged that one set of internalist norms is better than another if the first is more reliable in producing true beliefs.[116]

Internalist theories make the justifiability of a belief a function of the internal states of the believer, in the sense that if we vary anything but his internal states the justifiability of the belief does not vary. Thus the only properties of and relations between internal states to which internalist norms can appeal are those that cannot be varied without varying the internal states themselves. In other words, it must be necessarily true that if we are in those states then they have those properties and stand in those relations to one another. In short, they are "logical" properties of and "logical" relations between internal states. For instance, if S_1 is the state of believing $(P \mathbin{\&} Q)$ and S_2 is the state of believing P, then S_1 and S_2 are necessarily related by the fact that being in S_1 involves believing a

---

116. As Alvin Goldman does in his (1986).

conjunction whose first conjunct is believed if one is in state S_2. Thus we can characterize internalist theories as those proposing epistemic norms that appeal only to logical properties of and logical relations between internal states of the believer.

So, both internalism and norm externalism endorse internalist norms, but they differ in that, by definition, the internalist maintains that our epistemic norms are not subject to criticism on externalist grounds. It is hard to see how they could be subject to criticism on internalist grounds, so the internalist has typically assumed that our epistemic norms are immune from criticism—whatever our actual epistemic norms are, they are the correct epistemic norms. That, however, seems odd. On the surface, it seems it must be at least logically possible for two people to employ different epistemic norms. They could then hold the same belief under the same circumstances and on the basis of the same evidence and yet the first could be conforming to his norms and the second not conforming to his. If a person's epistemic norms are always beyond criticism, it would follow that the first person is justified in his beliefs and the second is not, despite the fact that their beliefs are based upon the same evidence. That would at least be peculiar. Because it seems that it must be possible for different people to employ different epistemic norms, this makes a strong prima facie case for norm externalism.

The prima facie case for norm externalism is bolstered when we notice that procedural norms are not generally immune to criticism. Typically, procedural norms tell us how to do one thing *by* doing something else.[117] For example, knowing how to ride a bicycle consists of knowing what more basic actions to perform—leg movements, arm movements, and the like—by doing which we ride the bicycle. An action that is performed by doing something else is a *nonbasic* action. Norms describing how to perform nonbasic actions can be subject to external evaluation. There may be more than one way to perform the nonbasic action, and some ways may be better (more efficient, more reliable, and so on) than others. If I know how to do it in one way and you know how to do it in another way, you know how to do it better than I if the norms governing your behavior are better than the norms governing mine. For example, we may both know how to hit the target with a bow and arrow, but you may know how to do it more reliably than I.[118] It thus becomes an empirical question whether acting in accordance with a proposed norm will constitute your doing what you want to be doing and whether another norm might not be better.

Reasoning is not, strictly speaking, an action, but it is something we do, and we do it by doing other simpler things. We reason by adopting new beliefs and rejecting old beliefs under a variety of circumstances.

----

117. The *by*-relation is what Alvin Goldman (1976a) calls *level-generation*.
118. Alternatively, we may have the same norms but your physical skills make you better able to conform to them.

Our norms for reasoning tell us when it is permissible or impermissible to do this. It seems that the norms we actually employ should be subject to external criticism just like any other norms. The norm externalist proposes that we should scrutinize them and possibly replace them by other norms. Because of the direct accessibility problem, we cannot replace them by norms making explicit appeal to reliability, but what we might discover is that (1) under certain circumstances inferences licensed by our natural norms are unreliable, and (2) under certain circumstances inferences not licensed by our natural norms are highly reliable. The norm externalist proposes that we should then alter our epistemic norms, adopting new internalist norms allowing us to make the inferences described under (2) and prohibiting those described under (1).

Some care is required here. We must distinguish between two construals of the norm externalist's proposal. He might be telling us that when we *discover* old reasoning patterns to be unreliable or new reasoning patterns to be reliable then we should alter our norms and our reasoning accordingly. Alternatively, he might be telling us that if old patterns simply *are* unreliable and new patterns *are* reliable, independently of our knowing or believing that they are, then we should alter our reasoning. The first construal seems like an eminently reasonable proposal, and it is one that has been made explicitly by various externalists. For example, in discussing how reliabilist considerations bear on reasoning, Goldman (1981) writes:

> At the start a creature forms beliefs from automatic, preprogrammed doxastic processes. ... Once the creature distinguishes between more and less reliable belief-forming processes, it has taken the first step toward doxastic appraisal. ... The creature can also begin doxastic self-criticism, in which it proposes *regulative* principles to itself (p. 47).

But this involves a fundamental misconception. Our epistemic norms are not subject to criticism in this way. Particular instances of reasoning are subject to such criticism, and the criticism can dictate changes in that reasoning, but this does not lead to changes in our epistemic norms. This is because unlike other norms, our epistemic norms already accommodate criticism based on reliability. The point is twofold. First, discovering that certain kinds of inferences are unreliable under certain circumstances constitutes a defeater for those inferences and hence makes us unjustified in reasoning in that way, and this is entirely in accordance with our natural unmodified epistemic norms. For example, we discover that color vision is unreliable in dim lighting, and once we discover this we should cease to judge colors on that basis under those circumstances. But this does not require an alteration of our epistemic norms, because color vision only provides us with defeasible reasons for color judgments, and our discovery of unreliability constitutes a defeater for those reasons. This is entirely in accordance with the norms we already have. Second, discovering that some new inferences are reliable under certain circumstances provides us with justification for making those inferences under those circumstances, but this is licensed by the norms we already have. That is precisely what induction is all about. For example, I might discover that I am clairvoyant and certain kinds of "visions" provide reliable indications of what is about to happen. Once I make this discovery it becomes reasonable for me to base beliefs about the future on such visions. Again, this is entirely in accordance with the norms we already have and does not require us to alter those norms in any way. The general point is that the kinds of reliability considerations to which the norm externalist appeals can lead us to reason differently (refrain from some old inferences and make some new inferences), but this does not lead to any change in our epistemic norms. Our actual epistemic norms are self-correcting in that they involve a kind of built-in feedback having the result that the sort of external criticism that could lead to the modification of other procedural norms does not necessitate any modification of epistemic norms.

We have had several externalists respond to this objection by protesting that they do not see the point of distinguishing between considerations of reliability leading us to alter our reasoning and those considerations leading us to alter our norms. But if all the externalist means is that considerations of reliability can lead us to alter our reasoning, then he is not disagreeing with anyone. In particular, he is not disagreeing with paradigmatic internalists like Chisholm. Norm externalism becomes nothing but a pretentious statement of a platitude.

The alternative construal of norm externalism takes it to be telling us that if old patterns of reasoning are unreliable and new patterns are reliable, then regardless of whether we *know* these facts about reliability, we should not reason in accordance with the old patterns and we should reason in accordance with the new patterns. What could the point of this claim be? It cannot be taken as a recommendation about how to reason, because it is not a recommendation anyone could follow. We can only alter our reasoning in response to facts about reliability if we are apprised of those facts. However, normative judgments do not always have the force of recommendations. That is, they are not always intended to be action-guiding. This is connected with the distinction that is often made in ethics between subjective and objective senses of 'should'. To say that a person subjectively should do X is to say, roughly, that given what he believes (perhaps falsely) to be the case he has an obligation to do X. To say that he objectively should do X is to say, roughly, that if he were apprised of all the relevant facts then he would have an obligation to do X. Judgments about what a person subjectively should do can serve as recommendations, but judgments about what a person objectively should do can only serve as external evaluations having some purpose other than guiding behavior.[119] The subjective/objective distinction can

---

119. They may serve as recommendations in an indirect fashion by conveying to a

be regarded as a distinction between evaluating the person and evaluating her act. The subjective sense of 'should' has to do with moral responsibility, while the objective sense has to do with what act might best have been performed.

We can draw a similar subjective/objective distinction in epistemology. The epistemic analogue of moral responsibility is epistemic justification. A person is being "epistemically responsible" just in case her beliefs are justified. In other words, epistemic justification corresponds to *subjective* moral obligation. What determines whether a belief is justified is what else the epistemic agent *believes* about the world (and what other directly accessible states she is in)—not what is in fact true about the world. This seems to show that whatever considerations of de facto reliability may bear upon, it is not epistemic justification. They must instead bear upon the epistemic analogue of objective obligation. What is that analogue? There is one clear analogue—objective epistemic justification is a matter of what you should believe if you were apprised of all the relevant truths. But what you should believe if you were apprised of all the relevant truths is just *all the truths*. In other words, the epistemic analogue of objective justification is *truth*. There is nothing here to give solace to a norm externalist.

Goldman (1981) draws a somewhat different distinction between two senses of 'justified' in epistemology. He distinguishes between "theoretical" evaluations of reasoning and "regulative" evaluations (the latter being reason-guiding). He suggests that the theoretical sense of justification is the sense required for knowledge and that it is to be distinguished from the reason-guiding sense. He suggests further that his reliabilist theory concerns the theoretical sense. The proposal is that it is knowledge that provides the point of a norm externalist's evaluation of epistemic norms in terms of considerations of reliability unknown to the epistemic agent. We do not believe that, but even if it were true it would not affect our overall point. The sense of epistemic justification with which we are concerned in this book is the reason-guiding or procedural sense, and if it is acknowledged that norm externalism bears only upon another sense of justification then our main point has been conceded.

To summarize the discussion of externalism, one can be an externalist by being either a belief externalist or a norm externalist. These exhaust the ways in which externalist considerations might be brought to bear on our epistemic norms. The belief externalist tries to formulate epistemic norms directly in terms of externalist considerations, but it is impossible to construct procedural norms in this way. The norm externalist proposes instead to recommend changes in procedural norms on the basis of considerations of reliability. Norm externalism initially appeared compelling because it provided a way to preserve the internal and procedural nature of epistemic norms while still allowing for external assessment

---

person that there are relevant facts of which he is not apprised.

of those norms. Combined with internalism's apparent inability to make sense of the comparative evaluation of norms, it seemed that norm externalism was very promising. Unfortunately, norm externalism fails on its two most plausible construals. It either fails to provide anything that reasoning within the framework of internalism does not, or it reduces justification to truth. So, norm externalism must be rejected.

As far as we can see, externalism has nothing to contribute to the solution to traditional epistemological problems. Justified beliefs are those resulting from normatively correct reasoning. Consequently, any evaluation of the justifiedness of a belief must be reason-guiding and hence must be beyond the pale of externalism.

4.4 Epistemological Relativism and the Individuation of Concepts

The apparent failure of norm externalism leaves us with a puzzling problem. Internalists have typically assumed that whatever epistemic norms we actually employ are automatically correct. But that seems hard to reconcile with the seemingly obvious fact that it is at least logically possible for different people to employ different norms. Surely, if Smith and Jones believe P for the same reasons, they are either both justified or both unjustified. There is no room for their justification to be relative to idiosyncratic features of their psychology resulting in their employing different epistemic norms. This seems to imply that there is just one set of correct epistemic norms, and the norms a person actually employs may fail to be correct. This conclusion would be obvious if it were not for the fact that there is no apparent basis for criticizing a person's norms. That is precisely what norm externalism tries unsuccessfully to do. The reliabilist considerations to which the norm externalist appeals are the only plausible candidates for considerations of use in criticizing and correcting epistemic norms, and we have seen that our epistemic norms cannot be corrected in this way. Of course, I might criticize Jones' norms simply because they disagree with mine, but he could equally criticize mine because they disagree with his. Are we committed to a thorough-going epistemological relativism then? That is at least unpalatable.

4.4.1 Theories of Individuation

The solution to the problem of relativism can be found by turning to a different problem. This is the problem of how concepts are individuated. If it could be shown that people who employ different norms are also necessarily employing different concepts in their reasoning, the troubling possibility of relativism would be dispatched. This is because epistemological relativism maintains that people are using the *same* concepts to reason according to different norms. We aim to show that people using different norms are employing different concepts, and this requires a substantial detour through a theory of the individuation of concepts. The detour will, however, yield a considerable payoff. We will secure a way to avoid relativism about epistemic norms even in the wake of the failure of norm externalism.

To understand the nature of the problem of concept individuation, first consider an analogous problem—that of object individuation. A theory of object individuation is a theory of what makes a physical object the object that it is, and by virtue of what two different objects are different. The historically most popular theory of object individuation proposes to individuate objects in terms of spatio-temporal continuity. On this account, object x and object y are the same object just in case they occupy the same space at the same time. Whatever you may think about the truth of this claim, it is a substantive theory and it is attempting to tell us something nontrivial about physical objects.

4.4.2 Truth Conditions

Theories of the individuation of concepts are similar. They attempt to tell us when concept A is the same concept as concept B. The standard theory takes concepts to be individuated by their truth conditions. The claim of this theory is that what makes a concept the concept that it is are the conditions that must be satisfied for something to exemplify that concept. These conditions comprise its truth conditions. The precise content of the truth condition theory of concepts deserves closer inspection than it usually receives. There is one sense in which the truth condition theory of concepts is correct but also completely trivial and uninteresting. The truth condition of the concept *red* is the condition of *being red*, and the truth condition of the concept *blue* is the condition of *being blue*. The following is undeniable:

red = *blue* if and only if *being red* = *being blue*

but it is hardly illuminating. Rather than explaining the concepts, the truth conditions presuppose the concepts. We might just as well define the "identity condition" of a physical object to be the condition of *being that object* and then claim that physical objects are individuated by their identity conditions. That is about as unilluminating as a theory can be. Unlike the spatio-temporal continuity theory of object individuation, it does not make a substantive claim.

Typically, philosophical logicians slide back and forth between the vacuous claim that concepts are individuated by their truth conditions and the considerably more contentious claim that concepts can be informatively characterized by (and only by) giving truth condition analyses of them. A truth condition analysis of a concept is a *definition* of the concept—an informative statement of necessary and sufficient conditions for something to exemplify the concept. We think it is fair to say that many philosophical logicians do not clearly distinguish between the vacuous claim and the contentious claim, or at least take the vacuous claim to somehow directly support the contentious claim. But we see no reason to think there is any connection between the two claims.

The simplest objection to the truth-condition-analysis theory is that most concepts do not have the kind of definitions required by the logical theory of concepts. Analytic philosophy in the mid-twentieth century concerned itself almost exclusively with the search for such definitions, and if we can learn anything from that period it is that the search was largely in vain. It is a very rare concept that can be given an informative definition stating truth conditions. This may seem surprising in light of the fact that dictionaries purport to give definitions, and all of the concepts investigated by analytic philosophers have dictionary entries. However it is illuminating to actually consider such a dictionary definition. One dictionary we consulted defined "horse" as "a large four-legged animal, domesticated for carrying riders and hauling loads." Whatever this definition is, it is not a statement of logically necessary and sufficient conditions for being a horse. For example, a horse does not cease being a horse if it loses a leg in an accident, or if it has never been ridden and used for hauling loads. And if the conditions enumerated by this definition were sufficient for being a horse, then camels would be horses as well. One might suppose that the lexicographers who wrote this definition just did a poor job, but we defy the reader to find a better definition. The only conditions that seem logically necessary for being a horse are very general ones like "occupies space" and perhaps "living creature", but these are far from adequate to distinguish horses from other animals. The real lesson to be learned from this is that dictionary definitions are not statements of logically necessary and sufficient conditions. Whatever they are, they do not provide the kind of analyses required by the truth condition theory. The importance of this simple objection cannot be overemphasized. Most concepts do not have definitions in the philosophical sense of logically necessary and sufficient conditions. For reasons we find mysterious, many philosophers seem to just ignore this and go on pretending that some form of the truth condition theory of concepts is correct.

4.4.3 The Logical Theory of Concepts

There is another strand to this story. Traditionally, the only logical relations between concepts that were recognized by philosophers were entailment relations. Concepts, as "logical items", were supposed to be individuated by their logical properties, and it seemed that the only logical properties concepts possessed were those definable in terms of their entailment relations to other concepts. This generates the picture of a "logical space" of concepts, the identity of a concept being determined by its position in the space, and the latter being determined by its entailment relations to other concepts. The claim that concepts must have definitions is just a more specific version of this general picture—one alleging that the position of a concept in logical space is determined not just by one-way entailments but by two-way logical equivalences. Some version of this picture has been prevalent throughout much of twentieth century philosophy, and it still plays a prominent role in philosophical logic. We will call this general picture of the individuation of concepts *the logical theory of concepts.* It has often been either confused with or identified with the truth condition theory.

The logical theory of concepts is subject to a rather deep epistemological problem. In general, the logical theory cannot make sense of reasons. To see this, let us begin with defeasible reasons. The logical theory appears to lead directly to the impossibility of defeasible reasons. We assume that what makes something a good reason for holding a belief is a function of the content of the belief. If the content of the belief is determined by entailment relations, then those entailment relations must also determine what are good reasons for holding that belief. The only kinds of reasons that can be derived from entailment relations are reasons that are themselves entailments—conclusive reasons. Thus we are forced to the conclusion that all reasons must be entailments. But this must be wrong, because we have seen that many epistemological problems cannot be solved in terms of conclusive reasons. Justified belief makes essential appeal to defeasible reasoning.

We might try distinguishing between "formal reasons" that derive from principles of logic and apply equally to all concepts, and "substantive reasons" that are specific to individual concepts and reflect the contents of those concepts. The preceding argument is really only an argument that the logical theory of concepts is incompatible with there being non-conclusive substantive reasons. Thus we could render the logical theory of concepts compatible with defeasible reasoning if it could be maintained that all legitimate defeasible reasons are formal reasons. The only plausible way of defending this claim is to maintain that the only legitimate defeasible reasons are inductive reasons and to insist that inductive reasons are formal reasons. This is to take induction to be a species of logic. On this view, there are two kinds of logic—deductive and inductive—and each generates formal reasons that pertain to all concepts and hence need not be derivable from the contents of individual concepts. For example, a conjunction $(P \,\&\, Q)$ gives us a reason for believing its first conjunct P regardless of what P and Q are. Similarly, it was traditionally supposed that inductive reasons are formal reasons pertaining equally to all concepts. This absolves us from having to derive inductive defeasible reasons from the essential properties of the concepts to which the reasons apply.

Unfortunately, this attempt to render the logical theory of concepts compatible with induction fails. It was pointed out in chapter one that induction does not apply equally to all concepts. Inductive reasoning must be restricted to projectible concepts. There is no generally accepted theory of projectibility, but it is generally recognized that what makes a concept projectible is not in any sense a "formal" feature of it. The simplest argument for this was given long ago by Nelson Goodman (1955). Define:

x is *grue* if and only if either (1) x is green and first examined before the year 2000, or (2) x is blue and not first examined before the year 2000.

x is *bleen* if and only if either (1) x is blue and first examined before the year 2000, or (2) x is green and not first examined before the year 2000.

'Grue' and 'bleen' are not projectible. For example, if we now (prior to the year 2000) examine lots of emeralds and find that they are all green, that gives us an inductive reason for thinking that all emeralds are green. Our sample of green emeralds is also a sample of grue emeralds, so if 'grue' were projectible then our observations would also give us a reason for thinking that all emeralds are grue. These two conclusions together would entail the absurd consequence that there will be no emeralds first examined after the year 2000. It follows that 'grue' is not projectible. Now the thing to notice is that 'blue' and 'green' are definable in terms of 'grue' and 'bleen' in the precisely the same way 'grue' and 'bleen' were defined in terms of 'blue' and 'green':

x is green if and only if either (1) x is grue and first examined before the year 2000, or (2) x is bleen and not first examined before the year 2000.

x is blue if and only if either (1) x is bleen and first examined before the year 2000, or (2) x is grue and not first examined before the year 2000.

Thus the *formal* relationships between the pair 'blue', 'green' and the pair 'grue', 'bleen' are symmetrical, and hence we cannot distinguish the projectible from the nonprojectible by appealing only to formal properties of the concepts. Projectibility seems to have essentially to do with the content of the concepts. Therefore, any explanation for the existence of inductive defeasible reasons must make reference to the particular concepts to which the reasons apply, and hence, on the logical theory of concepts, inductive defeasible reasons become as mysterious as any other defeasible reasons.

There is of course the further point, defended earlier, that epistemology requires more defeasible reasons than just inductive ones. Thus even if inductive reasons had turned out to be formal reasons, that would not entirely solve the problem of the possibility of defeasible reasons.

The next thing to notice is that the logical theory of concepts makes conclusive reasons just as mysterious as defeasible reasons. This has generally been overlooked, but it is really rather obvious. Epistemologists have noted repeatedly that logical entailments do not always constitute reasons. Some entailments are conclusive reasons and others are not reasons at all. The latter is because P may entail Q without the connection between P and Q being at all obvious. For example, mathematicians have proven that the Axiom of Choice entails Zorn's Lemma. These are abstruse mathematical principles apparently dealing with quite different subject matters, and just looking at them one would not expect there to be any connection between them. If, without knowing about the entailment, one were so perverse as to believe Zorn's lemma on the basis of the Axiom of Choice, one would not be justified in this belief. Once the entailment is known, you can become justified in believing Zorn's Lemma *partly* by appeal to the Axiom of Choice, but your full reason for believing Zorn's Lemma will be the conjunction of the Axiom of Choice and the proposition that if the Axiom of Choice is true then Zorn's Lemma is true. You are believing Zorn's Lemma on the basis of this conjunction rather than just on the basis of the Axiom of Choice. You can never become justified in believing Zorn's Lemma on the basis of the Axiom of Choice alone, so the latter is not a reason for the former.

On the other hand, some entailments do provide reasons. If I justifiably believe both P and $(P \rightarrow Q)$, I *can* justifiably believe Q on the basis of these other two beliefs. In this case I do not have to believe Q on the basis of the more complicated belief:

$$P \text{ and } (P \rightarrow Q) \text{ and if } [P \ \& \ (P \rightarrow Q)] \text{ then } Q.$$

To suppose that each instance of reasoning in accordance with *modus ponens* must be reconstructed in this way would lead to an infinite regress.[120] Thus some entailments are conclusive reasons and others are not. But the logical theory of concepts gives us no way to make this distinction. It characterizes concepts in terms of their entailment relations to other concepts, but, *a fortiori*, all entailment relations are entailment relations. There is nothing about the entailment relations themselves that could make some of them reasons and others not. Thus conclusive reasons become just as mysterious as defeasible reasons on the logical theory of concepts. This seems to indicate pretty conclusively that the logical theory of concepts is wrong. There has to be more to concepts than entailment relations.

4.4.4 Rational Roles

To argue that the logical theory of concepts is wrong is not yet to say what is right. The theory we want to endorse in its place is the epistemological theory of concepts. This theory begins by noting that concepts are both logical and epistemological items. That is, concepts are the categories whose interrelationships are studied by logic, and they are also the categories in terms of which we think of the world. The interrelationships studied by logic can all be reduced to entailment relations. Thus logic need not take note of any other features of concepts. Logic can get along with a cruder picture of concepts than can epistemology.

---

120. This was apparently first noted by Lewis Carroll (1895).

But a complete account of concepts must accommodate both logic and epistemology. There is good reason to think that the role of concepts in epistemology is fundamental. Not all entailment relations are conclusive reasons, but it seems likely that all entailment relations derive from "simple" entailment relations, where the latter are just those that are conclusive reasons. Thus a theory of concepts adequate for epistemology will very likely be adequate for logic as well. The question then becomes, "What kind of theory of concepts is adequate for epistemology?"

In epistemology, the essential role of concepts is their role in reasoning. Concepts are the categories in terms of which we think of the world, and we think of the world by reasoning about it. This suggests that concepts are individuated by their role in reasoning. What makes a concept the concept that it is is the way we can use it in reasoning, and that is described by saying how it enters into various kinds of reasons, both conclusive and prima facie. Let us take the *rational role* of a concept to consist of (1) the reason-schemas (conclusive or defeasible) licensing an inference to the conclusion that something exemplifies it or exemplifies its negation, and (2) the reason-schemas licensing conclusions that can be justifiably drawn (conclusively or defeasibly) from the fact that something exemplifies the concept or exemplifies the negation of the concept.[121] We have encountered reason-schemas throughout this book, and we are able to use them to individuate perceptual concepts such as red. For instance,

"S appears red to me" is a defeasible reason for me to believe that S is red.

In our view, the concept red is individuated in part by the fact that a certain kind of belief is licensed by the defeasible reason-schema that applies to it (we will explore various other reason-schemas in chapter seven). Taken together, the reason-schemas we use to think about the world constitute our epistemic norms. Thus, the epistemic norms governing a concept are descriptive of its rational role.

Our proposal is that concepts are individuated by their rational roles. The essence of a concept is to have the rational role that it does. If this is right, the explanation for how there can be such things as defeasible reasons becomes trivial. Defeasible reasons are primitive constituents of the rational roles that characterize concepts. Defeasible reasons need not have an origin in something deeper about concepts, because there is nothing deeper. In an important sense, there is nothing to concepts over and above their rational role. To describe the rational role of a concept is to give an analysis of that concept, although not a truth condition analysis.[122]

---

121. We are leaving out some subtleties, as they are not particularly relevant for the problem of relativism that we are presently trying to solve. For a lengthier discussion of rational roles, see Pollock (1989), chapters four and five.

122. This view of concepts is reminiscent of the verification theories of the logical

It should be noted that the rational role account of concept individ-uation is only distantly related to proposals in the philosophy of mind that fall under the label of *conceptual role semantics*. These are theories that claim that the nature of the meaning of a thought consists in the (typically inferential) relations that the thought has to other thoughts.[123] It has been frequently asserted that this view is committed to meaning holism, in the sense that that two people who have different beliefs and draw distinct inferences based on those beliefs will have different thoughts.[124] Our rational role semantics focuses only on the reason-schemas that guide reasoning. The beliefs that provide the premises for the reasoning are irrelevant to rational roles.

We think it is undeniable that concepts are individuated by their rational roles, and not (at least in any non-vacuous way) by their truth conditions. But some further explanation for all of this is required. *Why* are concepts individuated in this way? We will shortly propose an answer to this question. For the moment, however, we will simply take it as established that concepts are individuated in this way. The importance of this theory of concepts for the matters at hand is that it lays to rest the spectre of epistemological relativism. Epistemological relativism is the view that (1) different people could have different epistemic norms that conflict in the sense that they lead to different assessments of the justi-fiedness of the same belief being held on the same basis, and (2) there is no way to choose between these norms. The epistemological theory of concepts enables us to escape any such relativism. Because concepts are individuated by their rational roles, it becomes impossible for people's epistemic norms to differ in a way that makes them conflict with one another. The epistemic norms a person employs in reasoning determine what concepts she is employing because they describe the rational roles of her concepts. If two people reason in accordance with different sets of

---

positivists. Pollock first defended a theory of this sort in his (1968), and in more detail in his (1974), although in those publications he talked about "justification conditions" rather than rational roles, and used the term a bit more narrowly. This view of concepts is also related to the somewhat cruder views expressed by Michael Dummet (1975) and (1976) and Hilary Putnam (1979) and (1984).

123. See Hartry Field (1977) and Ned Block (1986). Under pressure from Hilary Putnam's Twin Earth cases (1975), many conceptual role theorists have defended so-called 'two factor' versions of the view. An internal factor fixes the narrow content of a thought for use in psychological explanation, and an external factor—typically a theory of truth—fixes the wide content of a thought for resolving questions of reference.

124. The most vocal critics who pursue this line are Jerry Fodor and Ernest LePore (1992). See also Rob Cummins (1989) and (1996). Cummins comes closest to offering an objection that might be applied to our rational role account when he alleges that reason-schemas are to be revealed by psychological investigation and that psychology requires that concepts be fixed first (1996, 43). Our account of reason-schemas will evade this criticism if we offer a methodology for determining our reason-schemas that does not require psychological explanations. In the next chapter, we offer just such a methodology.

epistemic norms, all that follows is that they are employing different concepts. Thus it is impossible for two people to employ different epistemic norms in connection with the same concepts. Their conceptual frameworks are determined by their epistemic norms. Epistemological relativism is logically false.

We have argued that if two people use different epistemic norms, then they are employing different concepts and this foils relativism. It might be wondered, finally, whether or not people actually use different epistemic norms. We doubt that there really is any variation in epistemic norms from person to person.[125] We suspect that epistemic norms are species-specific, but this is an empirical question.[126] In order to resolve these issues, what seems to be required is a methodology for determining what our epistemic norms are. We turn to that problem in the next chapter.

5. Conclusions

The main purpose of this chapter has been to understand how epistemic norms function. They guide us in our cognition, and beliefs are justified just in case they are held in compliance with epistemic norms, but the way in which they guide us proves more difficult to understand than epistemologists have often supposed. Many epistemologists have been tempted by the intellectualist model, according to which we make explicit appeal to epistemic norms. But the intellectualist model could not be a correct theory of the way epistemic norms function, because it would lead to an infinite regress. In order to comply with an epistemic norm, we would have to have a justified belief to the effect that the norm makes a certain prescription in the present case, and that would require us to comply with another epistemic norm. The principal insight of the chapter is that epistemic norms function in the same general way as other procedural norms. They are descriptive of our procedural knowledge of how to cognize, articulating what we "know to do" in cognizing. As with any procedural norms, we do not always succeed in complying with them, so there is a competence/performance distinction in epistemology just as there is in linguistics. A theory of the content of our epistemic norms is not just a description of what we do when we cognize.

---

125. This possibility is one of the central concerns in Stephen Stich (1992).

126. The conclusion that if different people employ different epistemic norms then they employ different concepts may seem puzzling because it appears to make it inexplicable how such people could communicate with each other. Even if our conjecture regarding the species-specificity of norms is false, it need create no difficulty for communication. Pollock has argued at length that concepts play only an indirect role in communication. (Pollock's entire theory of language is developed in his (1982). A briefer sketch of the theory can be found in chapter two of Pollock (1984). The reader who is concerned with this question should consult those books.)

That would be a performance theory. It is instead a competence theory, describing the way we know how to cognize, whether we actually do it that way or not.

Because epistemic norms are internalized, they must be able to function without conscious monitoring. This has two important consequences. First, epistemic norms must appeal only to internal states. Second, epistemic norms can appeal to any internal states, not just beliefs. So we simultaneously have a refutation of belief externalism and an explanation for why the doxastic assumption fails.

Norm externalism alleges that epistemic norms can be evaluated in terms of external properties like their reliability. It turns out, however, that such evaluations are already built into our actual epistemic norms. As such, external considerations cannot mandate changes in our norms. Apparently our epistemic norms are beyond criticism. This seems initially puzzling, but it is explained by endorsing a rational role theory of concept individuation.

The preceding remarks explain how epistemic norms work, but they do not determine which epistemic norms are correct. We turn next, then, to the methodology for determining the correct norms and finally, in chapter seven, to the norms themselves.

6
EPISTEMOLOGY AND RATIONALITY

1. Epistemological Theories

Epistemology is about rational cognition and how we can know the various things we claim to know. This gives rise to investigations on several different levels. At the lowest level, philosophers investigate particular kinds of knowledge claims. Thus we find theories of perceptual knowledge, theories of induction, theories of our knowledge of other minds, theories of mathematical knowledge, and so forth. At an intermediate level, topics are investigated that pertain to all or most of the specific kinds of knowledge discussed at the lowest level. Theories of reasoning, both deductive and defeasible, occur at this level. At the highest level we find general epistemological theories that attempt to explain how justified belief in general is possible. At this level we encounter versions of foundationalism, coherentism, probabilism, reliabilism, and direct realism. The highest level theories can be regarded as theories about epistemic justification itself. However, the highest level theories can be viewed in two different ways. On the one hand, they can be regarded as descriptions of the overall structural relations that give rise to epistemic justification. They tell us how the various constituents of cognition fit together to give us our knowledge of the world. We will refer to these as *structural theories of epistemic justification*. On the other hand, high-level theories can also be proposed as to *why* epistemic justification has the general structure it does. These theories typically take the form of logical analyses of the concept of epistemic justification. We will refer to these as *analytic theories of epistemic justification*.

The distinction between structural and analytic theories of epistemic justification has often been appreciated only vaguely. This is really a distinction between high-level theories and higher level theories. The analytic theories are theories about what would make a structural theory true. Viewed in this light, it is natural to regard foundationalism, coherentism, and direct realism as structural theories, and reliabilism as an analytic theory. A reliabilist might, for example, endorse a foundationalist structural theory on the grounds that it is a (perhaps the only) reliable way of acquiring knowledge. This suggests that reliabilism and foundationalism have different targets and are not automatically incompatible.

On the other hand, foundationalists and coherentists (and also direct realists and probabilists) have often viewed their theories as analytic theories as well. To do that, they must insist that they are not just giving structural accounts of epistemic justification, but logical analyses as well.

The claim would be that not only are their theories true as structural accounts, but also that nothing further *makes* them true. The claim is that these structures are *constitutive of* the concept of epistemic justification—the concept is to be analyzed by giving a detailed account of how beliefs come to be justified rather than by giving some overarching principle (like reliabilism) that *selects* the particular constituents of the correct structural theory. In other words, epistemic justification is to be analyzed by enumerating the principles that give rise to it.

The main objection to treating foundationalist, coherentist, and direct realist theories as analytic theories is that they give at best piecemeal, ad hoc seeming, analyses. They characterize epistemic justification in terms of a general structure and a lot of diverse unrelated principles regarding particular kinds of reasoning (perception, induction, other minds, etc.) without any general account of what ties all these principles together. As such, they are at least inelegant, and cannot help but leave us wondering if there isn't something more to be said which would explain how this agglomeration of principles comes to be the correct agglomeration.

Probabilism and reliabilism have been touted as giving general and elegant characterizations of epistemic justification that escape from the ad-hoc-ness objection. However, we argued above that neither kind of theory really accomplishes what it claims to accomplish. We have concluded that these theories do not provide the missing account we seek for what unifies the diverse constituents of a correct structural theory of epistemic justification.

In the end, none of the familiar structural theories of epistemic justification keeps its promise of giving a unified account of epistemic justification, and we are left with no other unifying proposals. This is theoretically unsatisfactory. What is the origin of this complex structure of epistemic principles that gives rise to justified beliefs? We can gain insight on this question by thinking more carefully about the concept of epistemic justification itself. Here we must be careful. We have distinguished between procedural epistemic justification and a concept of justification that is more intimately connected with knowledge and the Gettier problem. Our concern here is with procedural epistemic justification. It seems undeniable that a correct structural theory of procedural epistemic justification is going to posit a complex structure in which is embedded a large number of seemingly unrelated epistemic principles governing reasoning about particular subject matters. What unifies the diverse constituents of this structure and makes them all part of the true theory of epistemic justification?

To make progress, we should locate this question in a broader perspective. Epistemic cognition is just one part of rational cognition. There is a traditional distinction between epistemic cognition (cognition about what to believe) and practical cognition (cognition about what to do). Rational agents are agents that cognize rationally, and correct epistemic cognition is a subspecies of rational cognition in general. So asking what makes certain epistemic procedures correct is a special case of the more general question what makes cognitive procedures rational. Procedural epistemology is part of a general theory of rationality. In addition to its epistemological elements, a theory of rationality will include at least an account of how goals are to be selected for adoption and how actions are to be selected for performance.

2. Human Rationality

Let us turn then to the question, "What makes rational procedures rational?" We can illuminate the concept of rationality by considering the way in which philosophers have traditionally tried to answer questions about how, rationally, to perform various cognitive tasks (e.g., reasoning inductively). The standard philosophical methodology has been to propose general principles, like the Nicod principle, or principles of Bayesian inference, or the hypothetico-deductive method, and test them by seeing how they apply to concrete examples. In order to test a principle by applying it to concrete examples, we must know how the example should come out. Thus, for example, Nelson Goodman (1955) demonstrated the need for a projectibility constraint in induction by contriving his "grue/bleen" example. The demonstration was conclusive because everyone who looks at the example agrees that it would not be rational to accept conclusions defended by reasoning inductively in ways that violate the projectibility constraint. But how do people know that? The standard answer is "philosophical intuition", but that is not much of answer. What is philosophical intuition?[127]

Compare this methodology—the *methodology of intuitions*—with the methodology employed by linguists studying grammaticality. Linguists try to construct general theories of grammar that will suffice to pick out all and only grammatical sentences. They do this by proposing theories and testing them against particular examples. It is a fact that proficient speakers of a language are able to make grammaticality judgments, judging that some utterances are grammatical and others are not. These grammaticality judgments provide the data for testing theories of grammaticality. It is equally undeniable that human cognizers are able to make judgments about whether, in specified circumstances, particular cognitive acts are rational or irrational. These rationality judgments provide the data for testing theories of rationality.

We can imagine a philosopher arguing that rationality has to do with concepts and logic, so our intuitions about rationality are a kind of platonic intuition of universals. But no one would be tempted to say the same thing about our intuitions of grammaticality. After all, the details of our language are determined by linguistic convention. Our language could

---

127. The papers delivered at a recent conference on this topic have been collected by Michael DePaul and William Ramsey (1998).

have been different than it is, and if it were, our grammatical intuitions would have been different too.

When we learn a language, we learn *how to do* various things. Knowledge of how to do something is procedural knowledge. For most tasks, there is more than one way to do them. Accordingly, we can learn to do them in different ways, and so have different procedural knowledge. But, as noted above, a general characteristic of procedural knowledge seems to be that once we have it, we can also judge more or less reliably whether we are conforming to it in particular cases. For example, once I have learned how to ride a bicycle, if I lean too far to one side (and thus put myself in danger of falling), I do not have to wait until I fall down to know that I am doing it wrong. I can detect my divergence from what I have learned and attempt to correct it before I fall. Similarly, it is very common for competent speakers of a language to make ungrammatical utterances. But if they reflect upon those utterances, they have the ability to recognize them as ungrammatical and correct them. We suggest that exactly the same thing is true of cognition. We have procedural knowledge for how to cognize, and that carries with it the ability to recognize divergences from that procedural knowledge. That is what our so-called "philosophical intuition" amounts to, at least in the theory of rationality.

Recall that performance theories are theories about how people in fact behave. A competence theory, on the other hand, attempts to articulate people's procedural knowledge for how to do various things. Because people do not always conform to their own procedural knowledge, competence and performance can diverge dramatically. Our proposal is that the theories of rationality that philosophers construct by appealing to their intuitions about rationality are best viewed as competence theories of cognition. That is, they are attempts to articulate the rules comprising our procedural knowledge for how to cognize. In the case of epistemic cognition, those rules are our epistemic norms. We have no direct access to those rules themselves, but because we can detect (with fair reliability) divergences from our procedural knowledge, we can tell in real or imagined cases whether it would be rational to draw various conclusions (i.e., whether doing so would conform to our procedural knowledge). We can use such knowledge about particular cases to confirm general principles about the content of our procedural knowledge.

Unlike linguistic knowledge, it seems pretty clear that large parts of our procedural knowledge of how to cognize are built into us rather than learned. This may be required as a matter of logic—it may prove impossible to get started in learning how to cognize unless we already know how to cognize to some extent. But even if that is not true, it is overwhelmingly likely that evolution has built into us knowledge of how to cognize so that we do not come into the world so epistemically vulnerable as we would otherwise be. This way, we at least know how to get started in learning abut the world. This is rather strongly confirmed by the overwhelming agreement untutored individuals exhibit in their procedural knowledge of how to cognize.[128] For example, psychological evidence indicates that everyone finds reasoning with modus ponens to be natural and reasoning with modus tollens to be initially unnatural.[129] This is a very robust result of empirical investigations of human reasoning. It is unlikely that this is something we have learned. Once we start supplementing our built-in procedural knowledge with learned principles, we are quick to embrace modus tollens as well.

Notice that the rules comprising our procedural knowledge of how to cognize cannot be viewed as mere generalizations about how we do cognize. That would make them descriptions of cognitive performance rather than cognitive competence. Instead, they are the rules that we, in some sense, "try" to conform to. They are the rules perceived divergence from which leads us to correct our cognitive performance to bring it into conformance.

Human rationality is composed of the principles comprising our built-in procedural knowledge of how to cognize. When we turn more specifically to epistemology, this is what unifies the diverse collection of principles that make up a correct structural theory of epistemic justification. They are unified simply by being among the principles that are built into our cognitive architecture as human beings. There need be no overarching general characterization that explains why those are the principles we use. Those are the principles comprising our cognitive architecture because we just happen to be built that way.

The proposal is then:

A theory of human rationality is a competence theory of human cognition.

Such a theory proceeds by eliciting the procedural norms governing human cognitive performance. When the norms govern epistemic cognition, they are epistemic norms. Procedural epistemic justification consists of holding beliefs in compliance with correct epistemic norms. We propose this as our analysis of epistemic justification:

A belief is justified if and only if it is held in compliance with the cognizer's epistemic norms.

In understanding this analysis we must distinguish between doing something in accordance with norms and doing it in compliance with the

---

128. This is not to say that everyone is equally good at cognizing. Cognitive performance varies dramatically. But insofar as people make cognitive mistakes, they can generally be brought to recognize them as such, suggesting that their underlying procedural knowledge is the same.

129. See P. Wason (1966) and P. Cheng and K. Holyoak (1985). *Modus ponens* is the inference rule licensing an inference from P and $(P \rightarrow Q)$ to Q. *Modus tollens* is the inference rule licensing the inference from $\sim Q$ and $(P \rightarrow Q)$ to $\sim P$.

norms. The analysis proceeds in terms of the latter. To say that you act in accordance with a norm is just to say that your behavior does not violate the norm. This is compatible with your doing it for some reason unrelated to the norm. To say that you act in compliance with the norm is to say not only that you act in accordance with the norm but also that your behavior is guided by the norm. Justification requires compliance—not just accordance.

In one familiar sense of naturalism, this is a naturalistic analysis of epistemic justification (we will offer a full discussion of naturalism in section four). Cognition consists of natural processes. It is something we know how to do. To say that we know how to do it is to say that it is governed by internalized norms. Our epistemic norms are, by definition, the norms that actually govern our cognition. This, we claim, is a naturalistic definition of "epistemic norm". Of course, we have not proposed an informative logical analysis of the governance process which forms the basis of these definitions, but that should not be expected. This is a natural process that we can observe in operation, not just in reasoning but in all cases of internalized procedural knowledge, and its nature can be clarified by psychological investigations. But it must be emphasized that the only clarification that can be expected here is empirical clarification. We can no more provide an informative logical analysis of the governance process than we can provide an informative logical analysis of electrons or magnetism. These are natural kinds and natural processes that we discover in the world, and their nature is revealed by empirical investigation—not logical analysis.

It has been objected that the analyses of epistemic justification proposed by existing epistemological theories are piecemeal and uninformative. We insist that our analysis escapes this objection. There is nothing piecemeal about the analysis of epistemic justification as compliance with the norms comprising our procedural knowledge for how to cognize. This analytic theory of epistemic justification gives a completely general and unified account of epistemic justification. On the other hand, the *structural* theories associated with this analytic theory will automatically be piecemeal. This is a consequence of the nature of procedural norms. Such norms instruct us to do various things under various circumstances and prohibit us from doing other things. These norms have to be rather specific because, as we saw above, they must take as input only features of the present circumstances that are directly accessible to our cognitive systems. This precludes the possibility of the norms appealing to sweeping general features of the circumstances (features such as the belief being produced by a reliable process). Compare the norms for bicycle riding. These are going to be very specific, including such things as, "If you feel yourself losing momentum then push harder on the pedal" and "If you think you are falling to the right then turn the handlebars to the right". Epistemic norms will be equally specific, telling us things like as "If something looks red to you and you have no reason for thinking it is not red then you are permitted to believe it is red". There is no more reason to think that we can combine all epistemic norms into one simple general formula than there is for thinking there is a single simple formula governing the use of the pedals, the handlebars, the brakes, and so on, in bicycle riding. Procedural norms cannot work that way.

It is illuminating to contrast this account of epistemic norms with more conventional internalist formulas. Internalists have been inclined to say instead that our epistemic norms describe the way we *actually reason*. This claim has played an important role in internalist epistemology, because it tells us how to find out what proper epistemic norms are—just examine the way we actually reason.[130] But this is at least misleading. We do not always reason correctly, and what epistemic norms describe is *correct* reasoning. We might similarly be inclined to say that our bike-riding norms describe the way we actually ride a bicycle, but even when we know how to ride a bicycle we sometimes make mistakes and fail to conform to our norms—I might be distracted and lose my balance. Thus we might more accurately say that our bike-riding norms describe the way we actually ride a bicycle when we do it correctly. This formulation, however, sounds vacuous. After all, riding a bicycle correctly or reasoning correctly just is to conform to the norms. This creates a real puzzle for traditional accounts of procedural norms. The puzzle is resolved by seeing how norms for doing something are connected with knowing how to do it. The best way to describe the connection between norms and actual behavior is to say, as we did above, that our bike-riding norms and our epistemic norms are the norms that *actually guide us* in riding bicycles and reasoning. This is similar, in a very important respect, to the more customary claim that our epistemic norms describe the way we actually reason. In each case, norms are to be elicited from what we actually do and not from some mysterious criterion, separate from our actual behavior, that tells us what we should do. But there is also an important difference between the present formulation and the traditional formulation. The present formulation does not take our reasoning behavior at face value.

3. Epistemological Methodology

This general account of epistemic norms and epistemological theories has important implications for epistemological methodology. Epistemological theories are supposed to give general accounts of "right reasoning"—that is, they purport to describe our epistemic norms. It is a contingent psychological fact that we have the norms we have. Equivalently, it is a contingent psychological fact that we employ the conceptual framework we actually employ. Does this mean that epistemological

---

130. Chisholm (1977) endorsed this under the label "critical cognitivism", and Pollock endorsed it in his (1974) and called it "descriptivism".

theories are contingent? This is a rather complicated question. The answer is, "Partly 'yes', and partly 'no'." Part of what we do in epistemology is to elicit our actual epistemic norms, and that really is a contingent matter. But our ultimate conclusions are to the effect that particular concepts have rational roles of certain sorts. The rational role of a concept is a necessary feature of that concept, so it seems that our ultimate conclusions are, if true, necessarily true. Let us take this a bit more slowly, looking at each step of what transpires in an epistemological analysis.

We begin with a question such as, "How are we justified in forming beliefs about the colors of objects?", that is, "What are the rational roles of color concepts?" We begin our investigation by trying to determine how we actually make such judgments. This is a matter of eliciting the epistemic norms we actually employ. That is a question about human psychology. But this does not mean that the best way to go about answering it is by performing laboratory experiments. To illustrate, consider a simpler case. Typing is an excellent example of something we learn to do automatically. When we learn to type we internalize norms telling us what to do and then we follow those norms automatically. Now suppose we want to describe those norms. Consider the question, "What finger do you use to type a 'w'?" We *could* try to answer that question by designing a laboratory experiment in which we observe people typing 'w's under a wide variety of circumstances, but that would be silly. There is a much easier way to do it. We can *imagine* typing a 'w' and observe what we do. Touch typists find themselves using their left ring finger. How can this work as a way of eliciting our norms? After all, we are not just asking what finger a person uses on a particular occasion, and people do not always type correctly. What we want to know is what finger our typing norms prescribe using to type a 'w'. The reason we can answer this question by performing our thought experiment is that there is an introspectible difference between complying with one's internalized norms and not complying. We could perversely type the 'w' with our right index finger, but if we did we would know that we were not doing it the way we learned to do it. The explanation for the introspectibility of this difference is something we have already observed. Namely, it is required in order for procedural norms to be able to correct ongoing behavior. Thus it is important to the operation of procedural norms that compliance with them be introspectible.

Now consider how we can answer an epistemological question like, "How do we judge that something is red?", where this is intended to be a question about our epistemic norms. Sometimes we reflect upon actual judgments we observe ourselves making. More often we *imagine* making such judgments under normal circumstances and see what goes on. For example, suppose we are considering the hypothesis that something's looking red to us gives us a defeasible reason for thinking it is red. We imagine being in situations in which things look red to us and note that if there are no "intervening" considerations we will come to believe that the object is red. This is not just an observation about what actually happens. It is an observation about what we know *to do* in judging colors, that is, an observation about how our automatic processing system actually guides us in reasoning about colors. It is the introspectibility of complying with a norm that makes this observation possible.

This illustrates what goes on in epistemological analysis. Our basic data concern what inferences we would or would not be permitted to make under various circumstances, real or imaginary. This data concerns individual cases and our task as epistemologists is to construct a general theory that accommodates it. Epistemologists have often supposed that our epistemic rules should be, in some sense, self-evident.[131] We have been arguing that many of the individual bits of data on which our epistemological theory is founded will, in a certain sense, be self-evident (more accurately, introspectible). By virtue of knowing how to reason we know how to tell right reasoning when we see it, and that provides us with our data. But that does not guarantee that it will be easy to construct theories describing our epistemic norms or that such theories will be obviously right once we have them. One complication both in the use of thought experiments and in interpreting our data is that because our automatic processing system operates in a non-intellectual way without any conscious monitoring, it need not be obvious to us what makes a particular belief justified even when it is evident to us that it is justified. Our data consists in the fact that various beliefs *are* justified—not *why* they are justified. This can be illustrated by reflecting upon the fact that we have a much better account of perceptual knowledge than we do of many other kinds of knowledge. We have urged that our being appeared to in various ways provides us with prima facie justification for holding beliefs about our physical surroundings. The defense of this claim assumes that our beliefs in normal perception arise psychologically from our being appeared to in various ways. This is a contingent psychological thesis and cannot be regarded as a self-evident philosophical datum. Nevertheless, we regard it as a well established psychological fact, and so have no misgivings about assuming it in constructing an account of our epistemic norms.

Contrast epistemological theories of perceptual knowledge with those of a priori knowledge. We have no very good theories of a priori knowledge despite the fact that we have no difficulty telling which beliefs are justified and which are not when we are actually doing mathematics or logic. In other words, we know how to proceed in a priori reasoning, and hence we have the same kind of basic data as in the case of perception—we can recognize some beliefs as justified and others as not. What we lack in the case of a priori knowledge is a psychological account of what is going on when we have justified beliefs. We do not know the psychological source of such beliefs, and this hamstrings us in the attempt to construct theories of justification. This illustrates both the way in which our basic

---

131. This is what Ernest Sosa (1981) calls "methodism".

epistemological data are self-evident and the importance of contingent non-self-evident psychological facts in the construction of epistemological theories. In an important sense, describing our actual epistemic norms is part of psychology. This does not mean that it is best carried out in the laboratory, but neither can it be denied that the results of laboratory investigations can be relevant.

The contingent enterprise of describing our actual epistemic norms is not all there is to epistemology. From a description of our epistemic norms, we want to draw conclusions about the rational roles of various concepts, and that is a matter of conceptual analysis. But conceptual analysis is supposed to provide us with necessary truths. How is it possible to derive necessary truths from contingent psychological generalizations? In order to answer this question, a brief digression into philosophical logic is in order.

Conceptual analyses describe necessary properties of concepts. But notice that true statements about the necessary properties of things need not be necessarily true. To take a well-worn example, nine is the number of planets, and nine is necessarily such that it is odd, so it follows that the number of planets is necessarily such that it is odd; but the latter is only contingently true. This is because the necessity involved is *de re* rather than *de dicto*. Similarly, a statement describing the necessary properties of a concept must refer to the concept in some way, and if the mode of reference is only contingently a way of referring to that particular concept then even though the property ascribed to the concept is a necessary property of the concept, the resulting statement will be contingent. Applying this to epistemology, in describing epistemic norms we are describing necessary properties of concepts, but this does not mean that our epistemological pronouncements are themselves necessary truths. It depends upon how we are thinking of the concepts. For example, we might be thinking of the concept *red* under some description such as, "what is ordinarily expressed by the word 'red' in English". The meaning of an English word is a contingent matter, and so the claim that the concept *red*, so conceived, has such-and-such a rational role, will be a contingent claim about necessary properties of concepts.

Although conceptual analyses need not be expressed by necessary truths, there will be necessary truths lurking in the wings. We can think of propositions and concepts in terms of contingent descriptions of them. E.g., I may think of a proposition as "the first proposition entertained by Bertrand Russell on the morning of April 7, 1912". It is a contingent matter what proposition this is. But we also think of propositions and concepts in terms of their contents. E.g., we may think of a proposition as "the proposition that P". This is a noncontingent way of thinking of propositions and concepts. The description "the proposition that P" could not have picked out a different proposition than it does. If you think about a concept or proposition in this direct fashion and you ascribe a necessary property to it, then your belief is necessarily true. A conceptual analysis describes necessary properties of concepts, so if the conceptual analysis is expressed by a proposition that is about the concept directly then that proposition is necessarily true. Thus conceptual analyses do generate necessary truths. But they are not a priori truths. The analyses describe the rational roles of concepts, and our knowledge of those rational roles is derived from the discovery of contingent psychological generalizations regarding what epistemic norms we employ in reasoning. Thus the ultimate issue of epistemology is necessary *a posteriori* conceptual analyses.

4. Naturalized Epistemology

One of the aspirations of externalism was to provide an account of justification that fits into a picture of human beings as biological information processors. This picture views human beings as natural cognitive machines that evolved in response to environmental pressures and whose capacities are oriented toward achieving stability in a changing world. Some may find this conception of humankind depressing or pessimistic. On the contrary, contemporary philosophers tend to consider it a virtue when their views can be made consistent with the impressive advances of knowledge about how human beings work. Thus, there has been an effort to make epistemology *naturalistic.*

The promise of a substantial naturalistic epistemology was part of the attraction of externalism, as externalist views have seemed to be the only way to incorporate scientific information about biology and cognition into a theory of justification. We have, however, argued that this is a mistaken impression. Internalists can espouse a theory of justification that is just as naturalistic as externalist theories purport to be. We have been claiming that ours is a naturalistic epistemology, and it seems clearly naturalistic in at least one sense of naturalism. Since we have rejected externalism but still wish to offer an epistemology that is naturalistic, we owe a discussion of naturalism. Apart from externalism, assessing the urge toward naturalism has become an integral part of the domain of contemporary theories of justification, so it will be useful to spend a moment on the topic.

Epistemic naturalists are committed to making sure that epistemology is grounded in facts of nature and is responsive to insights gleaned from outside of philosophy. This characterization of naturalism, though, is far too vague to be of much use. What is needed is a more precise statement of what it means to ground epistemology in facts of nature. One can be an ontological naturalist or a methodological naturalist.

4.1 Ontological Naturalism

The most straightforward way of construing naturalism in epistemology is as the claim that epistemology ought to be naturalistic in its ontology. One is an ontological naturalist if epistemic terms such as knowledge or justification can in principle be analyzed in terms of natural entities or in terms of properties of natural entities. What is special about natural entities is that they are not themselves terms of epistemic evaluation, so it has been thought that an analysis of epistemic terms in the idiom of natural facts will yield a satisfying explanation of the nature of epistemic terms.[132] Ontological naturalism maintains that beliefs have their epistemic status because they have specifiable non-epistemic properties. These properties are to be facts of nature, and are to be embedded in our best understanding of the universe.

In spite of many philosopher's readiness to accept this as an important notion of naturalism, and in spite of the widespread endorsement of ontological naturalism in epistemology, ontological naturalism is not the principal naturalism in epistemology. Thinking of naturalism as primarily ontological is initially seductive, however, as it is invited by the comparison to naturalism in ethics. In ethics, ontological naturalism is the primary sense of naturalism, but the comparison between epistemology and ethics is misleading. Unlike ethics, where ontological *non*-naturalism is a viable position, epistemology claims few ontological non-naturalists. That is why it is not maximally illuminating to draw the contrast between naturalism and non-naturalism in epistemology at the level of ontology.[133]

4.2 Methodological Naturalism

A more useful way of understanding naturalism in epistemology emphasizes the need for philosophers to incorporate the fruits of a naturalistic methodology into their proposals, or to themselves employ a naturalistic methodology when framing their theories of justification. Methodological naturalism is not first and foremost a claim about the ontology of epistemology, though particular methodologically naturalistic strategies will often suggest an ontological commitment. In contrast with focusing on ontological naturalism, categorizing contemporary proposals in epistemology as methodologically naturalistic or non-naturalistic will reveal a landscape that is much more consonant with epistemologist's own judgments as to their naturalistic or non-naturalistic commitments.

It is not a simple matter to say which kinds of methodologies are naturalistic, or which sorts of results should be incorporated into epistemology. In its most common manifestation, the naturalistic project has been thought to involve incorporating the methods or results of science into epistemic matters. Judging from some of the recent literature on epistemic naturalism, naturalism and a robust commitment to science go hand in hand simply as a matter of definition.[134] The reasoning is that

---

132. Alvin Goldman defends this approach at the beginning of his (1979).

133. This point is made by Richard Foley (1994) and Alvin Goldman (1994). Alvin Plantinga (1993b) may be one of the rare examples of an epistemologist who is an ontological non-naturalist, as his Theory of Proper Function relies on the intentions of a divine designer.

134. Examples include Barry Stroud (1981), James Maffie (1995), and Richard

science offers the best resources for empirical inquiry into the details about those aspects of human beings that epistemologists are interested in. Not everyone, however, agrees with this pro-science attitude in understanding naturalized epistemology. Though they have been somewhat less influential in discussions of naturalism, there are a number of dissenters from the view that treats naturalism as automatically generating an interest in science.[135] In general, these philosophers pursue a less committal version of naturalism where the influence of empirical insight is understood widely enough to include many types of *a posteriori* inquiry. Susan Haack illuminates this more expansive view of naturalism:

> Does science have a special epistemic status? Thinking about this question at a commonsense level, unalloyed by any sophisticated epistemological theory, I should be inclined to answer 'yes and no'. 'Yes', because science has had spectacular successes, has come up with deep, broad and detailed explanatory hypotheses which are anchored by observation and which interlock surprisingly well with each other; 'no', because although, in virtue of those successes, science as a whole has acquired a certain epistemic authority in the eyes of the lay public, there is no reason to think that it is in possession of a special method of inquiry unavailable to historians or detectives or the rest of us, nor that it is immune from the susceptibility to fad and fashion, politics and propaganda, partiality and power-seeking to which all human cognitive activity is prone. (137)

So, there are epistemologists who identify methodological naturalism with a commitment to science, and those who have a more modest view of naturalism. It is possible that all this goes to show is that two camps in contemporary philosophy have appropriated the term 'naturalism' and that they apply it to two related but different projects. On the other hand, there may be a deeper motivation behind these two kinds of naturalism that can be identified. What the two camps seem to have in common is that they wish to use the results of *a posteriori* methodologies to influence their epistemological theories. This way of thinking about methodological naturalism includes the desire to bring scientific insight into epistemology. It also covers the more modest understanding of naturalism that is willing to draw widely from the social sciences or other types of empirically rich research.[136] We are led to the following definition of naturalism for a theory of justification.

Fumerton (1994).

135. For instance, Susan Haack (1993), Keith Lehrer (1990), and Jaegwon Kim (1988).

136. For instance, some feminist epistemologies that are avowedly naturalistic criticize epistemology for looking too quickly to science for empirical support. After adopting methodological naturalism it is a matter for further philosophical discussion just which methods are best. Science does not receive an automatic endorsement. Vrinda Dalmiya and Linda Alcoff (1993) explore the use of folk wisdom in naturalized epistemology, and Lorraine Nelson (1993) appeals primarily to feminist sociology.

DEFINITION:
A theory of justification is naturalistic if it maintains that epistemology should either consist partly or wholly in empirical disciplines, or should be informed by the results of empirical disciplines.

This definition covers a wide swath. There is no shortage of theories
that are methodologically naturalistic, even though the details of the
views vary widely.[137] Now that we have before us what naturalism is,
we must investigate the motivations behind methodological naturalism.
    Our general definition of naturalism includes as a possibility that
epistemology might consist wholly in empirical disciplines. This would
be a considerable departure from the philosophical enterprise of studying
knowledge and justification. The view that epistemology should consist
in the empirical study of perception, memory and induction is most
associated with Willard Van Orman Quine (1969), who is generally cred-
ited with having started the move toward naturalism in epistemology.
Quine advocates a complete assimilation of epistemology by psychology.
On the most obvious reading of his work, he argues that the *a priori*
methodology of philosophy will no longer have any role to play in epistemology. He concludes this from the failure of Cartesian-style doxastic
foundationalism. Quine's view appears to leave no room for normative
evaluations in epistemology. We have been presuming that a crucial
component of a theory of justification is to explain how we ought to
reason, so we view Quine's proposal as too extreme.[138]
    In the wake of Quine's work, epistemologists have tried to construct
a less radical naturalistic epistemology that acknowledges the crucial
role of normativity. There have been numerous attempts to incorporate
naturalism into a normative theory of justification. One influential way
is to acknowledge that the normative evaluation of beliefs is a philosophical
enterprise, but to restrict the range of acceptable theories to include only
the ones that are compatible with the results of empirical research. The
sense of "compatible" at work here needs some explaining. The way
empirical data may play a role in epistemology is to treat scientific results
as a negative constraint.[139] This is the view that an epistemological theory
offered alongside the best results of empirical research should not make
impossible demands on the cognizer. For example, Christopher Cherniak
(1986) has argued that assessing the logical consistency of a large corpus
of beliefs is too arduous for human cognizers or for any other real-time
computational device. We are invited to consider a device that detects
consistency by constructing a truth-table. For *n* atomic letters (each

---

137. See, for example, Louise Antony (1992), Alvin Goldman (1979), (1986) and
(1992), Gilbert Harman (1973) and (1986), Hilary Kornblith (1989) and (1993), James
Maffie (1995), Pollock (1995), W. V. O. Quine (1969), and James Taylor (1990).
    138. Also see Lawrence Bonjour (1994) and Jaegwon Kim (1988) for discussions of
this point.
    139. Christopher Cherniak (1986), Hilary Kornblith (1989), and Paul Thagard (1982).

corresponding to a believed sentence), the truth table will be 2^n lines long. Cherniak goes on to say,

> How large a belief set could an ideal computer check for consistency in this way? Suppose that each line of the truth table for the conjunction of all these beliefs could be checked in the time a light ray takes to traverse the diameter of a proton, an appropriate "supercycle" time, and suppose that the computer was permitted to run for twenty billion years, the estimated time from the "big bang" dawn of the universe to the present. A belief system containing only 138 logically independent propositions would overwhelm the time resources of this supermachine. Given the difficulties in individuating beliefs, it is not easy to estimate the number of atomic propositions that in a typical human belief system, but 138 seems much too low. (93-4)

One hundred and thirty-eight atomic propositions believed does indeed seem much too small to capture the contents of a healthy mind, but $2^{138} = 3.5 \times 10^{41}$. This leads to Cherniak's startling observation. Although coherence theories are sometimes accused of this failing, it is doubtful that a theory of justification that demands assessing the logical consistency of all of one's beliefs has ever been seriously proposed. Still, an obvious lesson to draw from Cherniak's insight is that a theory of justification that requires that we assess the logical consistency of a massive doxastic corpus will always lead to skepticism because none of the beliefs will ever be justified. The general message is that, in order to avoid skepticism, a theory of justification should not disregard known limitations of human reasoners. At work here appears to be a desire to keep the normative demands on an epistemic agent consonant with that agent's capabilities. In a word, this type of naturalized theorizing claims that epistemology should make *realistic* demands on human cognizers. We will call this the "realistic principle." This thesis is sometimes held to be part of a constraint on wider normative theorizing (for instance, in ethics or in philosophical discussions of practical reasoning) to the effect that an *ought* can only be normative if the agent is able to follow its requirements.[140] The realistic principle seems to be on the right track. It is unintuitive to many philosophers that a normative theory should make unrealistic demands.

In spite of the intuitiveness of the realistic principle, though, there have been objections to it in the epistemological literature. Richard Feldman and Earl Connee (1985) are skeptical of the realistic principle. They discuss the evidentialist conception of epistemology. The evidentialist view is, roughly, that a belief is justified if and only if it fits the evidence that the agent has. Feldman and Connee argue that this is the fundamental notion of justification. One kind of objection their view faces is that a bald evidentialism does not seem to take into account the cognitive limits of human beings. Sometimes epistemic agents are not able, due to the limits of their cognitive capacities, to come to beliefs based on all of their

---

140. Colloquially speaking, "ought implies can."

evidence. Feldman and Connee's response is essentially to deny that epistemic theories must make realistic demands on cognizers. They write,

> There is no basis for the premise that what is epistemically justified must be restricted to feasible doxastic alternatives. ...Suppose that there were occasions when forming the attitude that best fits a person's evidence was beyond normal cognitive limits. This would still be the attitude *justified* by the person's evidence. If a person had normal abilities, then he would be in the unfortunate position of being unable to do what is justified according to the standard for justification asserted by [evidentialism]. This is not a flaw in the account of justification. Some standards are met only by going beyond normal human limits. (19)

Feldman and Connee compare epistemic standards with the standards for getting an "A" in a difficult course, or the standards of excellence in art. It very well might be, they urge, that such standards are beyond the capabilities of normal human beings. By analogy, Feldman and Connee are suggesting that some normative standards in epistemology are defensible even though they may entail that few people are able to meet those standards.

We are faced with two conceptions of the limitations on normative theorizing in epistemology in particular and in philosophy in general. Any conclusion we come to would have wide implication for debates in philosophy given the ubiquity of normative issues, so a cautious approach is in order. In some sense Feldman and Connee are surely right. Normative standards in general should not be restricted to a range that many or most human beings are routinely able to attain. This seems especially compelling in their case of artistic excellence. Not everyone is constitutionally able to be a great artist. Thinking otherwise would diminish the artistic achievement of those who are held in artistic esteem. On the other hand, there does seem to be something suspicious about a normative standard that no one is able to attain, or that is attainable in only the most extraordinary of circumstances. The way these two extremes usually get manifested in epistemology tracks our prima facie response to skepticism. Recall from chapter one that a few epistemologists view the primary philosophical challenge of a theory of knowledge or justification as defeating the skeptic. They think that skepticism is a live possibility that needs to be specifically dispatched. If the skeptic should remain unbeaten, something important will be revealed about the human intellectual condition. Epistemologists who think that skepticism is a tolerable conclusion to philosophical theorizing tend to allow that a theory of justification might be so demanding as to make most of our beliefs unjustified.[141] Of course, there is the threat here that this skeptical conclusion itself might be unjustified so that there would be no reason to accept it, so care will need to be taken in order to avoid self-refuting versions of skepticism.

---

141. Keith Lehrer (1990) vigorously defends the possibility of skepticism.

168 CHAPTER SIX

We have been supposing that skepticism must be rejected from the start and that the most progress in epistemology will be made by focusing squarely on how it is possible that at least some, and probably many, of our beliefs are justified. Epistemologists who think that skepticism is intolerable will tend to desire that a theory of justification be within the capabilities of at least most cognizers.

We suspect that there is a no defensible formula for determining when a limitation of an epistemic agent that is revealed by empirical inquiry should be treated as evidence for skepticism versus those occasions when it should invite modifications to the epistemological theory. There is a continuum of cases, and it is not clear whether we can say anything general about all of them. The proper reply, then, to Feldman and Connee is that, if they wish to avoid skepticism, they would do well to have *some* degree of sensitivity to the actual capabilities of epistemic agents. We have before us a reasonable prima facie case for seeking epistemological theories that are realistic in their demands on human cognizers. This case follows the diminishing fortunes of the possibility that a thorough-going skepticism in epistemology is a tenable result of philosophical inquiry into human intellectual flourishing.

This reply also shows a way to address some of the most well-entrenched sources of resistance to naturalized epistemology. Naturalists in epistemology are commonly criticized for having to bootstrap their theories. The complaint is that if science (or any other a posteriori inquiry) is used to inform debates in epistemology, it will be an open and troubling question as to how the naturalistic epistemologist will be able to assess the credentials of the science she is appealing to. Laurence Bonjour (1985) summarizes this worry about naturalism when he writes,

> Since what is at issue here is the metajustification of an overall standard of empirical knowledge, rather than merely an account of some particular region of empirical knowledge, it seems clear that no empirical premises can be employed. Any empirical premise employed in such an argument would have to be either (1) unjustified, (2) justified by an obviously circular appeal to the very standard in question, or (3) justified by appeal to some other standard of empirical justification (thereby implicitly abandoning the claim that the standard in question is the correct overall account of epistemic justified for empirical beliefs). Thus the argument would apparently have to be purely *a priori* in character. (10)

Bonjour is claiming that it would be circular to secure the credentials of an empirical enterprise based on a theory that contains empirical commitments, some of which may have come from the empirical enterprise in question. Arguments of this form usually amount to assertions of epistemology as first philosophy, and they lead to the conclusion that epistemology must be wholly *a priori*. [142] Responses to this problem are

---

142. This argument can be found in numerous other places, including Bonjour (1994), Roderick Chisholm (1989), Jonathan Dancy (1985), Keith Lehrer (1990), and

extensions of Quine's claim that we have no choice but to bootstrap ourselves to an acceptable epistemological theory using the resources of a posteriori inquiry. This will be an appropriate response to the critic of naturalism only if there is a good reason not to put all of human inquiry under epistemological scrutiny at the same time. The main motivation behind doubting all of human knowledge in order to conduct epistemology is skepticism. But if global skepticism is inappropriate, then epistemological theorizing is just one more aspect of our general attempt to understand the world, and there is no reason not to use any relevant knowledge we may have in the course of that endeavor. In other words, we know how to engage in epistemic endeavors, and our procedural knowledge of how to do that does not depend upon any prior epistemological theory. The search for an epistemological theory is not an attempt to *justify* human cognition, but to *describe* it. It doesn't need prior justification.[143]

It would be a mistake to portray methodological naturalism as the dominant attitude in contemporary epistemology, even though it underwrites the dominant naturalistic attitude in contemporary epistemology. One of the driving forces behind methodological *non*-naturalism is that the reply of the last paragraph seems too simple. To many philosophers, it seems illegitimate to reject skepticism and then to declare that some areas of human knowledge do not need to be questioned. In response, note that the alleged infelicity of naturalism does not seem much worse than the difficulty with using an a priori methodology to conduct epistemology and then to defend that methodology by appeal to a priori arguments. The non-naturalist who hopes to retain a priori reasoning as the primary resource for epistemology will apparently have to provide some defense other than a priori reasoning to secure the credentials of her methodology or risk running into the same circularity that the naturalist does.[144]

So far, we have been discussing the idea of a general kind of negative constraint on epistemology in order to keep theories of justification realistic. The view that an epistemological theory should not make unrealistic demands may draw from the fund of evidence found in a range of empirical inquiry. We have yet to make one more step to illuminate the most common version of naturalism in epistemology. Rather than remaining neutral on the types of empirical data that must be accommodated to keep a theory of justification realistic, philosophers have deemed that cognitive psychology is especially worthy of attention.[145] This is because

---

Michael Williams (1991).

143. Related formulations of this reply can be found in Ronald Giere (1985) and Philip Kitcher (1992).

144. Hilary Kornblith (1994a) develops this line. Laurence Bonjour, at least, is sensitive to this issue. See especially his (1998, p. 99).

145. For a detailed historical overview, see Philip Kitcher's (1992) landmark review of epistemological naturalism. Hilary Kornblith's introduction to his (1994b) also contains an illuminating discussion of the relationship between psychology and naturalized

cognitive psychology is commonly taken to include the rigorous study of how human beings come to form beliefs, the very subject matter of epistemology.

4.3 Naturalism and Psychology

The close relation between the subject matter of cognitive psychology and of a theory of justification has provided some impetus to narrow the scope of the realistic principle. To see how this is so, recall that earlier we argued that we must keep separate the reasons that are evidence for a belief in some purely logical or epistemological sense and reasons that are the *cause* of a particular belief in a particular believer. Our claim was that this distinction was necessary to preserve the distinction between justified and justifiable belief. Several prominent naturalists in epistemology have offered a slightly different argument to this same conclusion.[146] They have concluded from this insight that we must appeal to whatever empirical research program takes as one of its targets the understanding of the causal relations between beliefs and other mental states. This research program is cognitive psychology and, to some extent, cognitive science. Naturalists think that traditional, non-naturalized epistemology has mistakenly satisfied itself with giving a description of correct evidential relations. The problem, as they view it, is that epistemology is not solely concerned with the theory of correct evidential relations. It is also aimed at the theory of the conditions under which actual cognizers realize those relations in their thinking.

The issue in some ways harkens back to the very first considerations that motivated epistemology. Where epistemology is viewed as having the task of describing human intellectual flourishing, a theory of justification might simply name all the reasonings that would yield an ideal intellectual agent, whether or not anyone ever realized those reasonings. This is the strategy that naturalists object to. An alternate possibility is that a theory of justification would emphasize the connection between the reasonings of an ideal agent and the reasonings that actual agents realize. On the second view, it would not do to simply list all the truths that an intellectually flourishing agent could endorse, nor would it do to name all the logical and evidential relationships between truth-conducive statements. Naturalists are interested in the actual psychological transactions that obtain between mental states. In Alvin Goldman's (1986) theory, for instance, the connection between epistemology and cognitive psychology dovetails with the framework of reliabilism. Once he has defended the relevance of psychology and has argued for reliabilism, he appeals to

---

epistemology.
146. Most notably, Gilbert Harman (1973), Alvin Goldman (1979), (1985) and (1986), and Hilary Kornblith (1980).

the data from psychology to determine which cognitive processes are most reliable. The theory that results is, of course, externalist. It also incorporates the details of the naturalistic science of the mind.

We are now in a position to recapitulate the elements of naturalism in contemporary epistemology. Epistemologists have been led to methodological naturalism based on a particular picture of the world and the epistemic agent's place it in. This picture has it that epistemic achievements are part of the physico-causal structure of the mind and the world, as opposed to merely the consequence of logical or abstract evidential relations between propositions. If this picture is correct, then it would stand to reason that a non-epistemological methodology would be available to study that structure. As we have seen, this insight is really just a consequence of taking seriously the partly causal character of the basing relation. These arguments for naturalism are neutral on which sort of empirical inquiry will provide the most illuminating resources for epistemology. Given the goals of naturalism, it should be an open question as to which sort of empirical insight will ultimately prove the most helpful. Be that as it may, many epistemologists with naturalistic convictions have staked a claim on which empirical research we should attend to. Psychology has seemed to be the best candidate for investigating the structure of the relations between beliefs and other mental states. Furthermore, the best known versions of reliabilism have a proprietary way of incorporating psychological results. Unsurprisingly, reliabilist theories are interested in the reliability of the cognitive processes implicated in belief formation.

4.4 Naturalistic Internalism

Naturalism has been prominently associated with externalism, but the resources for the type of arguments that motivate naturalism are not unique to externalism. Externalist theories provide one way of filling out the relationship between naturalism and epistemology, once the drive toward naturalism is defended, but internalist theories can endorse naturalism, too. Since the crucial underlying insight in rejecting the purely evidential conception of epistemology is not unique to externalism, it can be incorporated into an internalist theory. The theory of epistemic norms defended in the last chapter is simultaneously internalist and naturalistic. It proposes that epistemic norms are "psychologically real" features of human cognition. They represent contingent features of the human cognitive architecture. What is novel about our naturalistic internalism is that it does not carry the commitment that the best way to investigate epistemic norms is by employing the standard methodologies of empirical psychology. Those methodologies are well suited for pursuing performance theories of cognition, but not for pursuing competence theories. That is, laboratory investigations of what people do under various circumstances yield, *a fortiori*, theories of what people do under those circumstances. It is a difficult jump to move from that to a theory of the content of the procedural knowledge that leads them to do whatever they do. At this stage of the development of cognitive psychology, the most appropriate methodology for investigating procedural knowledge is the standard philosophical methodology. The philosopher constructs thought experiments, makes judgments about whether certain kinds of beliefs or inferences would be epistemically justified in the circumstances envisioned, and then tries to construct a theory capturing the resulting "philosophical intuitions". It is noteworthy that this is also the favored methodology of the linguist studying grammaticality. In both epistemology and linguistics, the subject matter is something psychologically real, but the best way to study it may not be with the standard tools of contemporary cognitive psychology.

Although cognitive psychology is currently ill equipped to investigate epistemic norms, naturalistic information can be relevant to epistemological theorizing in another way, reminiscent of some of the points made by Cherniak. Although standard scientific investigations may have difficulty telling us what the contents of our epistemic norms are, they may often be able to tell us what they are *not*. Epistemic norms formulate procedural knowledge, and as such it must be possible to conform to them. If for either psychological or computational reasons, human cognizers could not comply with a proposed norm, it follows that that norm is not part of their procedural knowledge for how to cognize, and hence that it is not one of the procedural norms comprising human rationality. This strategy is remarkably effective in eliminating some versions of theories that have been quite popular among philosophers.[147] For example, a common view has it that in choosing what actions to perform, a rational agent should consider all possible actions and select one that maximizes expected utility. But this is clearly impossible. There are always infinitely many possible actions. No one can consider them all, compute their expected utilities, and then select an optimal one. Such a norm cannot be complied with. It follows that it is not one of the norms comprising human rationality. Some similar more heavily qualified norm probably is, but this particular norm has been stated too simply to be a viable candidate for a principle of human rationality.

The preceding considerations suggest that a useful way of investigating human rationality, and more specifically epistemology, is to approach it from what Daniel Dennett (1971) calls "the design stance". Consider how it would be possible to build a humanlike cognitive agent having the kinds of cognitive capabilities that human beings exhibit. This has the potential to illuminate not only on *what* the contents of human epistemic

---

147. Louise Antony (1992) offers a careful and effective discussion of this use of psychology in epistemology.

norms are, but also *why* they have the content they do. It can be argued that many of the general features of human cognition represent the only, or the only obvious, way of performing various cognitive tasks consistent with the various logical and computational constraints that must be satisfied by any real agent.[148] The next section will be devoted to a brief discussion of this approach to illuminating the structure of rational cognition.

5. Generic Rationality

Theories of human rationality are specifically about human cognition. As competence theories of human cognition, they describe contingent psychological features of the human cognitive architecture. Rational agents (artificial or otherwise) can have a more general structure that differs from human cognition in significant ways, but still be the sorts of things we consider rational. To illustrate this with a simple example, it was noted above that there is overwhelming psychological evidence that human beings do not employ *modus tollens* as a primitive inference rule. They can learn to use it, but it is not built into their system of cognition from the start. If we could construct an artificial agent that was the cognitive duplicate of human beings except that it also employed *modus tollens* as a primitive inference rule, we would not regard its cognitive behavior as irrational, despite the fact that it would not conform exactly to human norms of rational cognition. What is this more general notion of rationality?

Although a theory of human rationality is concerned with describing contingent features of the human cognitive architecture, it is not entirely an accident that humans are built the way they are. Environmental pressures have led to our evolving in particular ways. We represent one solution to various engineering problems that were solved by evolution. There is no reason to think that these problems always have a single, or even a single best, solution. Certain engineering problems call for arbitrary choices between solutions that work equally well. Our cognitive architecture probably reflects a great many decisions that would be made arbitrarily by an engineer designing a cognitive agent like us. On the other hand, many of the more general features of our cognitive architecture may reflect the only, or one of a small number of, solutions to general problems of cognitive engineering. For example, it can be argued that very general logical and computational constraints on cognition dictate that any sophisticated cognitive agent will engage in defeasible reasoning, will reason defeasibly from perceptual input, will reason inductively, and will engage in certain kinds of planning behavior.[149] When that is

---

148. This is one of the main theses defended in Pollock (1995).
149. See Pollock (1995).

true, it provides an informative explanation for why human cognition works as it does.

Approaching rationality from the design stance, we propose to understand generic rationality as attaching to any agent that constitutes a solution to the design problem that also generates the human cognitive architecture. What is this design problem that rationality is designed to solve? We will begin by exploring one possibility, and then generalize the account.

A simplistic view of evolution has it that evolutionary pressures select for traits that enhance survivability of the creature. So we might regard the design problem for rationality to be that of creating an agent that can survive in a hostile world *by virtue of its cognitive capabilities*. This, however, presupposes a prior understanding of what cognition is and how it might contribute to survivability. We take it as characteristic of rational cognition that a cognitive agent has doxastic states ("beliefs", broadly construed) reflecting the state of its environment, and conative states evaluating the environment as represented by the agent's beliefs. It is also equipped with cognitive mechanisms whereby it uses it beliefs and conations to select actions aimed at making the world more to its liking. This constitutes the *doxastic-conative loop*, as diagrammed in figure 6.1. A rational agent has beliefs reflecting the state of its environment, and it likes or dislikes its situation. When it finds the world not entirely to its liking, it tries to change that. Its *cognitive architecture* is the mechanism whereby it chooses courses of action aimed at making the world more to its liking. Within the doxastic-conative loop, epistemic cognition is, in an important sense, subservient to practical cognition. The principal function of cognition is to direct activity (practical cognition), and the role of epistemic cognition in rationality is to provide the factual background required for practical cognition.

We take it, partly as a definition, that rational agents implement the doxastic-conative loop. We can evaluate how well they implement it. A judgment of how well an implementation performs must always be relative to a set of design goals that one implementation may achieve better than another. As remarked above, a natural design goal is to construct an agent capable of using its cognitive capabilities to survive in an uncooperative environment. This is motivated by considerations of evolution and natural selection, but it is based upon a simplistic view of evolution. For biological agents, a more natural design goal is propagation of the genome. And for artificial agents, we may have many different design goals. For example, in designing a selfless robotic soldier, the objective is neither survival nor propagation of the genome. This suggests that there is no privileged design goal in terms of which to evaluate cognitive architectures.

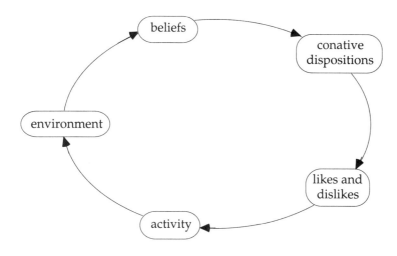

Figure 6.1   Doxastic-conative loop

It is striking, however, that for agents that implement the doxastic-conative loop, the same cognitive features seem to contribute similarly to achievement of a wide range of design goals. For example, an agent operating in a complex and uncooperative environment and intended to achieve its design goal by implementing the doxastic-conative loop will probably work better if it has a rather sophisticated system of mental representation, is capable of both deductive and defeasible reasoning, and can engage in long-range planning. This suggests that there is a relatively neutral perspective from which we can evaluate cognitive architectures, and this in turn generates a generic concept of rationality.

6. Truth and the Evaluation of Cognitive Architectures

The human cognitive architecture is what it is, and human rationality is defined relative to it. However, generic rationality allows a broad range of cognitive architectures differing in at least some respects. We can evaluate these cognitive architectures in terms of how well they achieve their design goal. This is reminiscent of norm externalism's evaluation of epistemic norms in terms of reliability. But one big difference is that practical and epistemic cognition are evaluated as a package. The ultimate objective is not truth, but practical success. Still, it seems that the production of true beliefs ought to be at least indirectly valuable in achieving practical success. Is this a way of resurrecting reliabilism?

The pursuit of true beliefs could be involved in the assessment of a cognitive architecture in several different ways. First, it might be proposed that we want an agent to believe as many of the truths as possible. However, that could be achieved by simply believing everything. Clearly, that would not contribute to practical success. Second, it might be proposed that we want an agent to have *only* true beliefs. That, however, could be achieved by believing nothing. Again, that will not contribute to practical success. An initially more plausible idea is that we want the agent to simultaneously maximize true belief and minimize false belief, or better, we want the agent to maximize the ratio of true beliefs to false beliefs. But this could be achieved by believing every tautology of the form (P ∨ ~P) and nothing else. Again, this will not contribute to practical success.

The last example illustrates that it is not truth per se, but *interesting* truth that the agent should be pursuing. Interesting truths are those that are particularly relevant to deciding how to act so as to achieve one's goals. But even here, it is not clear to what extent truth is the relevant desideratum. The agent should pursue beliefs that will help it be effective in achieving its goals, and in some cases it seems clear that truth is required. For example, it seems desirable to have true beliefs about whether a tiger is about to eat you. But in other cases it is not so clear that true beliefs will be the most helpful. True beliefs might be too complex to be used efficiently in day to day deliberation. Approximately true beliefs that are simple may be more useful. Consider deciding what route to take while walking across your front lawn to your door. If you had to do that by solving a problem in quantum mechanics, you would have a terrible time getting into your house. It is much more useful to have a number of approximately true "rules of thumb" that you can use in making such decisions, e.g., "If you try to walk over the tricycle rather than around it you are apt to fall down."

The upshot is that it is not clear how the pursuit of true belief enters into the evaluation of cognitive architectures. It seems clear that truth is often a good thing, but not all truths are equally desirable, nor are all falsehoods equally undesirable. The relationship of truth to rationality is not a simple matter. In particular, any attempt to resurrect reliabilism in this way is doomed to failure.[150]

7. How to Build a Person

We have spent considerable effort describing how epistemic norms are related to human rationality. In chapter seven we will return to the question of describing the content of these norms, defending a more precise version of direct realism. But in the meantime, even without going into detail, we may be left wondering why human epistemic norms

---

150. Stephen Stich comes to much the same conclusion in his (1992).

(and rational norms more generally) have the general structure they do. A great deal of progress on this question can be made by approaching it from the design stance and asking how we could build a cognitive agent with humanlike cognitive abilities. This enterprise motivates Pollock's *OSCAR Project*, which began in the mid-eighties. The goal of the OSCAR Project is the construction of a general theory of rationality and its implementation in an artificial rational agent. The implemented system will be an artificial intellect (an *artilect*) of the sort constructed in AI, but where it differs from most AI systems is that will be based upon a detailed philosophical theory of rationality. The computer implementation of philosophical theories represents a marked divergence from conventional philosophical methodology. The armchair-trained philosopher may wonder why we should bother with implementation. Of course, the simple answer is that it is fun to build something that actually works. But the more serious answer is that implementing an abstract theory of rational cognition is the only way to be sure that it actually will work. The lesson that any nascent computer programmer learns at her PC's knee is that programs almost never do what you expect them to do the first time around. Writing a computer program that actually does what you want is a matter of making a first attempt and then repeatedly testing it and refining it. Applying this lesson to philosophical theories of rational cognition, a theory that looks good from your armchair will almost never work the way you expect if you write a program that directly implements your first thoughts on the matter.

Implementation achieves two things. First, it requires the theorist to be precise and to think the details through. Philosophers are much too prone to ignore the details, just waving their hands when the going gets rough. That might be all right if the details were *mere* details and we could be confident that filling them in was a matter of grunt work. But in fact, when philosophical theories fail it is usually because the details cannot be made to work. Grand pictures painted with broad brushstrokes are fine for hanging on the wall and admiring for aesthetic reasons, but if the objective is to discover truth, it is essential to see whether the details can be made to work. So the first thing implementation achieves is that it requires the theory to be sufficiently precise that it can actually be implemented. It is remarkably common when implementing a theory to discover to your chagrin that there are significant parts of the theory that you simply overlooked and forgot to construct. To the armchair-bound philosopher, that may sound remarkably stupid, but that is only because he has never tried implementing his theories and has thereby never had the opportunity to make the same humbling discoveries about his own thought.

The second thing that implementation achieves is that it provides a test of correctness for theories of cognition. A theory of cognition is a theory of how to achieve certain cognitive tasks. Only by implementing the theory can we be certain that the theorized procedures do indeed accomplish their objectives. The armchair bound philosopher attempts to do this by looking for counterexamples. That is, in effect, what implementation is doing as well. Implementation enables us to use the computer as a tool in searching for counterexamples because it allows us to apply the implemented theory to concrete examples. But this technique far outstrips what can be accomplished searching for counterexamples while firmly implanted in your armchair. The difficulty with the latter is twofold. First, as remarked above, you may simply be wrong about the consequences your theory has for a specific example. The theory may not work the way you expect it to. If it is implemented, you can apply it to the example mechanically and not be misled by your own expectations. Second, the examples to which you can apply your theory from the armchair are severely limited in their complexity. Truly complicated examples are simply beyond our ability to work through them in our heads. But if the theories are implemented, they can be applied mechanically to a much broader range of more complex examples. The armchair philosopher, in his naivete, may suppose that if a theory is going to fail, it will fail on simple examples. That is just not true. AI has a long history (long for AI at least) of constructing theories and testing them on "toy examples". For example, much early AI work on planning was tested on the blocks world, which is a world consisting of a table top with children's blocks scattered about and piled on top of each other, and the planning problems were problems of achieving certain configurations of blocks. AI learned the hard way that systems that worked well for such toy problems frequently failed to scale up to problems of realistic complexity. There is every reason to expect the same thing to be true of philosophical theories of rational cognition. The only way to give them a fair test is to implement them and apply them to problems of real-world complexity.

Implementing a theory of rationality can show that it works, in the sense that a cognitive agent described by the theory can perform the cognitive tasks that the implemented system successfully performs. This does not, however, establish that human rational cognition works in the manner of the theory. There may be more than one way to carry out the cognitive task. Nevertheless, the implementation can provide indirect evidence for the claim that human rational cognition works in the manner described. As remarked above, a necessary condition for an account of rational norms to correct describe human cognitive competence is that it be possible for a cognitive agent to actually work that way. That turns out to be a difficult condition to satisfy. It is hard to construct *any* system capable of performing sophisticated cognitive tasks. Consequently, if there is independent reason to think that human cognition works in a certain rather general way, constructing a precise system conforming to the general description and showing that the system actually works provides some evidence for the claim that human cognition works that way, or at least in a very similar way.

Let us turn then to the task of building a cognitive agent with humanlike cognitive capabilities. We will take it one step at a time, beginning with simple cognitive abilities and then making them more complex. Our objective is to design an "intelligent machine" that could interact with its surroundings, learn from experience, and survive in a reasonably hostile environment. Let's call our machine 'Oscar'. What would we have to put into Oscar to make him work? At the very least we would have to provide him with ways of sensing the environment and thinking about the world. It is worth pursuing some of the details.

7.1 Oscar I

We must begin by incorporating sensors much like our sense organs so that Oscar can respond to states of the environment, and we might even call these sensors "sense organs". We must also incorporate "reasoning" facilities, both deductive and inductive. And we must incorporate some sort of conative structure to provide goals for Oscar to attempt to realize. If Oscar is to survive in a hostile environment, it would also be wise to provide sensors that respond to conditions under which he is in imminent danger of damage or destruction. We might call these "pain sensors". Oscar could then have built-in "fight or flight" responses elicited by the activation of his pain sensors.

We have described Oscar as thinking about the world. That involves a system of mental representation—what we might call a "language of thought".[151] For Oscar to have a thought is for him to "entertain" a sentence in his language of thought and treat it in a certain way. Without going into details, we can suppose abstractly that for Oscar to have a thought is for him to have a sentence in his language of thought residing in his "B-box".[152] Adopting a computer metaphor, we can think of the latter as a memory location. Oscar's thoughts and beliefs must be causally related to his environment and his behavior. On the one hand, Oscar must be constructed in such a way that the stimulation of his sensory apparatus tends to cause him to acquire certain beliefs. On the other hand, Oscar's having appropriate beliefs must tend to cause him to behave in corresponding ways. To describe these causal connections will be to describe his systems of epistemic and practical cognition.

Let us call the machine resulting from this stage of design "Oscar I".[153]

7.2 Oscar II

Oscar I could function reasonably well in a congenial environment. But in an environment that is both reasonably complex and reasonably hostile, Oscar I would be doomed to early destruction. He would be easy meat for wily machinivores. The difficulty is this. To be effective

---

151. This term comes from Jerry Fodor (1975).
152. The "B-box" metaphor is due to Stephen Schiffer (1981).
153. Oscar I is pretty much the same as the machines discussed by Hilary Putnam (1960).

in avoiding damage, Oscar I must not only be able to *respond* to the stimulation of his pain-sensors when that occurs—he must also be able to *predict* when that is apt to occur and avoid getting into such situations. He must be able to exercise "foresight". As we have constructed him, Oscar I has the ability to form generalizations about his environment as sensed by his sense organs, but he has no way to form generalizations about the circumstances in which his pain-sensors are apt to be activated. This is because Oscar I has no direct way of knowing when his pain-sensors are activated—he has no way of "feeling pain". As we have described them, the pain-sensors cause behavioral responses directly and do not provide input to Oscar's cognitive machinery. If Oscar is to be able to avoid pain rather than merely respond to it, he must be able to tell when he is in pain and be able to form generalizations about pain. To do this he needs another kind of sensor—a "pain-sensor sensor" that detects when the pain-sensors are activated. (Of course, the pain-sensors can themselves be pain-sensor sensors if they send their outputs to more than one place. We do not need a separate organ to sense the operation of the first organ.) Suppose we build these into Oscar I, renaming him Oscar II. This gives him a rudimentary kind of self-awareness. If the conative structure of Oscar II is such that he is moved to avoid not only the current activation of his pain-sensors but their anticipated activation as well, then this will enable him to avoid getting into situations that would otherwise result in his early demise.

It is illuminating to note that the difference between Oscar I and Oscar II is roughly the difference between an amoeba and a worm. Amoebas only *respond* to pain (or more conservatively, what we can regard as the activation of their pain-sensors)—worms can learn to avoid it. The learning powers of worms are pretty crude, proceeding entirely by simple forms of conditioning, but we have said nothing about Oscar that requires him to have greater learning powers.

Beginning with Oscar II we can distinguish between two kinds of sensors. First, Oscar II has *external sensors* to sense the world around him. These are of two kinds. He has ordinary perceptual sensors, and he also has pain-sensors that respond to environmental stimuli that tend to indicate impending damage to his body. Oscar II also has an *internal sensor* to sense the operation of his pain-sensors. His internal sensor could be described as a "higher-order sensor" because it senses the operation of another sensor.

The distinction between having a pain and feeling the pain is analogous to the distinction between perceiving the world by having a percept and being aware of the percept itself. Both reflect the operation of second-order sensors. It was observed that we are rarely aware of our percepts. We perceive the world "through" our percepts, but when we do it is the world we are aware of, not the percept. Pains, on the other hand, tend to grab our attention. We are rarely in pain without being aware of it. The aspect of the human cognitive architecture is probably a reflection of the fact that it is usually more important to note pains and form generalizations about what causes them than it is to note percepts and form generalizations about them.

7.3 Oscar III

Oscar II is still a pretty dumb brute. We have described him as sensing his physical environment and forming generalizations on that basis. But he does not do a very good job of that. The trouble is that he can only take his perception of the environment at face value. If his "red sensor" provides the input 'red' to his cognitive machinery, he can relate that to various generalizations he has formed concerning when there are red things about, and he can also use the input to form new generalizations. But the generalizations at which he will arrive will be crude affairs. He will have no conception of the environment fooling him. For example, he will be unable to distinguish between a machine-eating tiger and a mirror image of a machine-eating tiger. All he will be able to conclude is that some tigers are dangerous and others are not. We, on the other hand, know that all tigers are dangerous, but that sometimes there is no tiger there even though it looks to us like there is. Oscar II has no way of learning things like this. He has no way of discovering, for example, that his red sensor is not totally reliable. This is because, at least until he learns a lot about micromechanics, he has no way to even know that he has a red sensor or to know when that sensor is activated. He responds to the sensor in an automatic way, just as Oscar I responded to his pain-sensors in an automatic way. If Oscar II is to acquire a sophisticated view of his environment, he must be able to sense the activation of his red sensor.[154] That will enable him to discover inductively that his red sensor is sometimes activated in the absence of red objects.

This point really has to do with computing power and resource constraints. Given sufficient computing power, Oscar might be able to get by, forming all of his generalizations directly on the basis of the output of his external sensors. His generalizations would parallel the kind of "phenomenalistic generalizations" required by the phenomenalist epistemologies championed in the first half of this century by such philosophers as Rudolf Carnap, Nelson Goodman, and C. I. Lewis.[155] The most salient feature of such generalizations would be their extraordinary complexity. Just imagine what it would be like if instead of thinking about physical objects you had to keep track of the world entirely in terms of the way things appear to you and your generalizations about the world had to be formulated entirely in those terms. You could not do it. Human beings do not have the computational capacity required to form and confirm such complex generalizations or to guide their activities in terms of them. Instead, human beings take perceptual input to provide only defeasible reasons for conclusions about their physical environment. This allows

---

154. Hilary Putnam (1960) overlooks this.
155. See Rudolf Carnap (1967), Nelson Goodman (1951) and C. I. Lewis (1946).

them to split their generalizations into two parts. On the one hand they have generalizations about the relations between their perceptual inputs and the state of their environment, and on the other hand they have generalizations about regularities within the environment that persist independently of perception of the environment. The advantage of dividing things up in this way is that the two sets of generalizations can be adjusted in parallel to keep each manageably simple under circumstances in which purely phenomenalistic generalizations would be unmanageable. Epistemologically, we begin by trusting our senses and taking their pronouncements to be indicative of the state of the world. More formally, appearance provides us with defeasible reasons for judgments about the world and, initially, we have no defeaters for any of those judgments. Making initial judgments in this way we find that certain generalizations are approximately true. If (a) we can make those generalizations exactly true by adjusting some of our initial judgments about the world, and (b) we can do it in such a way that there are simple generalizations describing the circumstances under which things are not as they appear, we take that as a defeater for the initial perceptual judgments that we want to overturn and we embrace the two sets of generalizations (the generalizations about the environment and the generalizations about the circumstances under which perception is reliable). The result is a considerable simplification in the generalizations we accept and in terms of which we guide our activities.[156] A secondary effect is that once we acquire evidence that a generalization is approximately true, there is a "cognitive push" toward regarding it as exactly true.

The logical form of what goes on here is strikingly similar to traditional accounts of scientific theory formation. On those accounts we begin with a set of data and then we "posit theoretical entities" and construct generalizations about those entities with the objective of constructing a theory that makes correct predictions about new data. There is a formal parallel between this picture and our thought about physical objects. Physical objects play the role of theoretical entities, our sensory input provides the data, and we try to adjust the generalizations about physical objects and the "bridge rules" relating physical objects and sensory input, in such a way that we can make correct predictions about future sensory input. Of course, all of this is to over-intellectualize what goes on in human thought. We do not invent physical objects as theoretical entities

---

156. Philosophers of science have long been puzzled by the role of simplicity in scientific confirmation. When two theories would each explain the data but one is significantly simpler than the other, we take the simpler one to be confirmed. But this is puzzling. What has simplicity got to do with truth? The explanation for the role simplicity plays in confirmation may lie in the kinds of considerations we have been describing. Its importance has to do with minimizing computational complexity, and its legitimacy has to do with the fact that, in a sense, the objects the generalizations are about are "free floating" and can be adjusted to minimize complexity. This is a bit vague, but we find it suggestive. For related discussion, see Ronald Giere (1990) and (1991).

designed to explain our sensory inputs. We just naturally think in terms of physical objects, and our conceptual framework makes that epistemologically legitimate independent of any reconstruction of it in terms of scientific theory formation. Our point is merely that the logical structure is similar. From an information-processing point of view, the adoption of such a logical structure gives us an additional degree of freedom (the physical objects, or the theoretical entities) that can be adjusted to simplify the associated generalizations and thus minimize the computational complexity of using those generalizations to guide activity.[157]

The point of all this is that to acquire the kind of manageable generalizations about the environment that will enable him to keep functioning and achieve his built-in goals, an intelligent machine must be able to sense the operation of his own sensors. Only in that way can he treat the input from these sensors as defeasible and form generalizations about their reliability, and the need to treat them this way is dictated by considerations of computational complexity. Let's build such second-order sensors into Oscar II and rename him 'Oscar III'. He thus acquires a further degree of self-awareness. The difference between Oscar II and Oscar III may be roughly parallel to the difference between a bird and a cat. Kittens quickly learn about mirror images and come to ignore them, but birds will go on attacking their own reflections until they become exhausted.

Although Oscar III has second-order sensors sensing the operation of his first-order "perceptual" sensors, this does not mean that he can respond to his perceptual sensors only by sensing their operation—that is the mistake of the foundations theorist. In the ordinary course of events Oscar III can get along fine just responding mechanically to his perceptual sensors. To attend to the output of a sensor is to utilize (in cognition) the output of a higher order sensor that senses the output of the first sensor. Oscar III need not attend to the output of his perceptual sensors under most circumstances because doing so would not alter his behavior (except to slow him down and make him less efficient). He need attend only to the output of his second-order sensors under circumstances in which he has already discovered that his first-order sensors are sometimes unreliable. This is related to the fact that Oscar III will automatically have reason to believe that sense perception is generally reliable. (Of course, he might be wrong about this.)

The cognitive role of the pain-sensors is a bit different from that of the perceptual organs. Oscar III will function best if he almost always attends to the output of his pain-sensors. These play a different kind of role than the perceptual sensors. Their role is not just one of fine-tuning.

---

157. There is an interesting purely formal question here. That is the question of the extent to which and the circumstances under which computational complexity can be decreased by introducing such "intervening variables". It is obvious that this can sometimes be achieved, but it would be interesting to have a general account of it.

Except in "emergency situations" in which all cognitive powers are brought to bear to avoid a permanent systems crash, Oscar III should always be on the lookout for new generalizations about pain, and this requires that he almost always be aware of when his pain-sensors are activated. This parallels the fact that in human beings we are much more aware of our pains than of our visual sensations. We generally "look through" our visual sensations at the world and do not think about the sensations themselves.

We have attributed two kinds of self-awareness to Oscar III—he has the ability to sense the activation of his pain-sensors and also to sense the activation of his perceptual organs. These proceed via "internal" or "higher order" sensors. The important thing to realize is that there are simple explanations for why such self-awareness will make an intelligent machine work better. Other kinds of self-awareness may also be either desirable or necessary. We have not described Oscar III as having any awareness of what goes on internally after he acquires perceptual input or pain stimuli. In particular, we have not described him as having any way of sensing the operation of those cognitive processes whereby he forms generalizations on the basis of his perceptual inputs and pain stimuli. But such awareness seems to be required for two reasons. First, consider defeasible reasoning. In defeasible reasoning we reason to a conclusion, and then subsequent reasoning may lead to new conclusions that undercut the original reasoning and cause us to retract the original conclusion. In order for such negative feedback to work, the cognitive agent must be able to sense and keep track of his reasoning processes. Actually, humans are not terribly good at this. We forget our reasons rather rapidly, and we often fail to make appropriate corrections even when we remember our reasons.[158] We would probably work better if we could keep better track of our reasoning processes. At any rate, the general point seems clear. The ability to sense his own reasoning processes will be required in order for Oscar to indulge in defeasible reasoning, and defeasible reasoning seems to be required by any kind of sophisticated epistemology.

There is another reason a well-functioning cognitive agent must be able to sense his own reasoning processes to some extent. A cognitive agent does not try to gather information at random—he or she normally seeks to answer specific questions (motivated ultimately by conative considerations). Effective problem solving (at least in humans) involves the use of reasoning strategies rather than random permissible reasoning, and we acquire such strategies by learning about how best to search for solutions to various kinds of problems. To learn such things we must be

---

158. Recent psychologists have delighted in documenting human subjects' failure to make corrections to reasoning in light of new information. The first of these studies is apparently that of L. Ross, M. R. Lepper, and M. Hubbard (1975). For a thorough discussion of making corrections to reasoning about scientific theories, see Barbara Koslowski (1996).

aware of how we proceed in particular cases so that we can make generalizations about the efficacy of the search procedures employed.

7.4 Mental Representations

We described the Oscarites as having a language of thought to encode information. This language of thought is a representational system. It must, among other things, provide ways of thinking about particular objects and ascribing properties to them. We will call ways of thinking of objects "mental representations". Until the 1970's, philosophy adopted a rather parochial view of mental representations, often recognizing only one way of thinking of an object—as the unique object having a certain combination of properties. This is to think of an object "under a description". But as Saul Kripke (1972), Keith Donnellan (1972), and Hilary Putnam (1973) observed, this is actually one of the least common ways of thinking of objects. This is most easily seen by considering how often you can actually find a description that uniquely picks out an object without the description itself involving a way of thinking about another object. For instance, I can think of my mother as "the mother of me", but that only works insofar as I already have some way of thinking of myself. It is often fairly easy to propound descriptions that pick out objects uniquely as long as we are allowed to build into the descriptions relations to other objects. But to get such descriptions going in the first place, we must begin with some other ways of thinking of at least some objects. Those other ways could involve descriptions provided those descriptions did not make reference to further objects, but we challenge the reader to find even one such description. If we cannot find such descriptions, then it seems clear that they do not constitute the mental representations in terms of which we think of objects.

The unavoidable lesson to be learned from the paucity of descriptions is that we have nondescriptive ways of thinking of at least some objects. We do not have to look very far to find some of those nondescriptive mental representations. First, consider perception. When I see an object and make a judgment about it, I do not usually think of that object under a description—not even a description like "the object I am seeing" (I am typically seeing many different objects at any one time). Instead, I just focus my attention on the object and have a thought whose closest expression in English is something like "That is a table". In a case like this, my visual experience involves a visual representation of an object—a *percept*—and I think of the object in terms of that percept. That percept is my mental representation of the perceived object, and it is a constituent of my thought.[159] Percepts are not descriptions, so this is an example of a nondescriptive mental representation.

---

159. For related accounts of mental representation in perception, see Kent Bach (1982), Romane Clark (1973), Pollock (1982), and David Woodruff Smith (1984) and (1986).

A percept can only represent an object while that object is being perceived. If I later see another object that looks precisely the same way to me, then precisely the same percept will recur, but this time it will represent the new object. I can, however, continue to think about the original object after I am no longer perceiving it. When I do that I am no longer thinking of it in terms of the percept, so I must be employing a different kind of mental representation. This new mental representation still need not be a description. Once I have become able to think about an object in some way or other, I can continue to think about it even when that original mental representation is no longer available to me. This is clear in the case of objects originally represented by percepts, but it is equally true of objects originally thought about under descriptions. Thinking of an object under a description, I may acquire a wide variety of beliefs about it, but I may eventually forget the original description. For instance, I might have first come to think of Christopher Columbus under some description like "The man my teacher is talking about", but I can no longer remember just what description I might have used and I may no longer remember that Christopher Columbus satisfies that description. Such forgetfulness does not deprive me of my ability to think of Christopher Columbus. I have a nondescriptive way of thinking of Christopher Columbus. Such nondescriptive ways of thinking of an object are parasitic on originally having some other way of thinking of the object (either perceptual or descriptive), but they are distinct from those other ways. We call these nondescriptive ways of thinking of objects '*de re* representations', and Pollock written about them at length elsewhere.[160]

The above remarks are largely remarks about the phenomenology of human thought. They amount to the observation that we rarely think of objects under descriptions, but we do have both perceptual and nonperceptual nondescriptive ways of thinking of objects. As a remark about human psychology, this seems obviously correct, but it may also seem philosophically puzzling. How can there be such nondescriptive ways of thinking of objects? We can resolve this puzzle by reflecting on Oscar. Mental representations are just singular terms in the language of thought. If it can be shown that there is no obstacle to constructing Oscar in such a way that his language of thought contains such singular terms, then there should be no reason to be suspicious of the claim that human

---

160. See Pollock (1981) and (1982) (60ff). Those publications go into some detail in describing the workings of *de re* representations, but those details are largely irrelevant to the present discussion. Related discussions occur in Diana Ackerman's (1979), (1979a), and (1980). Pollock proposed that a number of Keith Donnellan's (1972) well known examples are best understood as illustrating the occurrence of *de re* representations in our thought.

beings employ such mental representations. From an information-processing point of view we can think of *de re* representations as pigeon holes (memory locations) into which we stuff properties as we acquire reasons to believe that the objects represented have those properties. Properties may drop out of the pigeon holes if they are not used occasionally (i.e., we forget). In order to establish a pigeon hole as representing a particular object we must begin by thinking of the object in some other way, and that initial way of thinking of the object will be the first thing to be put into the pigeon hole. For example, we might begin by thinking of an object under a description, from that acquire a *de re* representation of the object, and then we might eventually forget the original description and only be able to think of the object under the *de re* representation. From an information-processing point of view there are good reasons (having to do with the efficient use of memory) for incorporating something like *de re* representations into the language of thought used by an intelligent machine, and we would be well advised to equip Oscar with some such device.

Another important representational device is that involved in first-person beliefs. Numerous philosophers have observed that, although we can think of ourselves in more familiar ways (e.g., under descriptions, or perceptually), we can also have thoughts about ourselves in which we do not think of ourselves in any of those ways.[161] We will follow David Lewis in calling these *de se* beliefs.[162] The mental representations employed in *de se* beliefs will be called *de se* representations. The existence of *de se* representations is illustrated by the following example (due to John Perry 1979):

---

161. See H. N. Castaneda (1966), (1967), and (1968), Roderick Chisholm (1981a), David Lewis (1979), John Perry (1977) and (1979), Pollock (1981) and (1982) (13ff). This has also played a role in some recent work in artificial intelligence. See L. Creary and C. Pollard (1985), Rapaport (1984), and Rapaport and Shapiro (1984).

162. The existence of *de se* beliefs raises numerous philosophical questions about the analysis of what is believed. Castaneda and Pollock have both argued that in *de se* belief one believes a proposition containing a particular kind of designator—what Pollock previously called a "personal designator", but what might more aptly be called a "*de se* designator". Roderick Chisholm, David Lewis, and John Perry, on the other hand, have all urged that *de se* belief does not take a propositional object. They claim that the object of *de se* belief is instead a property or concept and *de se* belief involves a unique form of self-attribution. Fortunately, we need not get involved in this mare's nest at the moment. We doubt there is any substantive difference between the two accounts. Without some further constraints on what it is for something to be an object of belief, we can describe *de se* belief in either way. For example, Lewis' arguments to the contrary turn upon the assumption that any cognitive agent can, in principle, entertain any proposition. Pollock's endorsement of *de se* propositions led him to deny that and insist instead that some propositions are logically idiosynchratic. But there appears to be no way to substantiate either position without *first* resolving the question whether belief must take a propositional object.

> I once followed a trail of sugar on a supermarket floor, pushing my cart down the aisle on one side of a tall counter and back down the aisle on the other, seeking the shopper with the torn sack to tell him he was making a mess. With each trip around the counter, the trail became thicker. But I seemed unable to catch up. Finally, it dawned on me. I was the shopper I was trying to catch.

What happened when Perry realized that he was the shopper with the torn sack? He came to believe an identity, viz., that the shopper with the torn sack was the same person as he himself. This identity involves thinking of the same individual (himself) in two different ways (with two different mental representations) and believing that they *are* the same individual. The first mental representation is a straightforward descriptive representation—Perry thinks of himself as "the shopper with the torn sack". But the second representation, in which he thinks of himself as "me, myself", is a unique representation different from the kinds of representations we can employ in thinking of other things. I can think of myself under a description, or perceptually (e.g., I may see myself in a mirror), or in terms of a *de re* representation, but whenever I think of myself in one of those ways I may fail to realize that it is myself that I am thinking of. To realize that I am thinking of myself is to relate the mental representation I am employing to a special way of thinking of myself—my *de se* representation of myself.

Why do we have such a special way of thinking of ourselves? *De se* representations are essential elements of the language of thought of any sophisticated cognizer (human or otherwise). This is most easily illustrated by considering the conative aspects of Oscar. The purpose of providing Oscar with cognitive powers is to enable him to achieve built-in goals. This is accomplished by combining his beliefs with a conative structure consisting of preferences, desires, aversions, and so on, and a cognitive structure leading him to behave in specifiable ways in the presence of particular combinations of beliefs and conations. The problem now is to construct a cognitive structure of the latter sort in a creature lacking *de se* beliefs. We claim that it cannot be done. Practical reasoning consists of forming the intention to do something under specified circumstances, forming the belief that you are in such circumstances, and then performing the action. We must supply the cognitive agent with rules for the formation of appropriate intentions and beliefs. The intentions are *conditional* intentions to do something *if* some particular condition is satisfied. To be useful, the conditions must concern the agent's situation and not just the general state of the universe. They must involve the *agent's* being in the specified circumstances. The rules for the formation of such conditional intentions must be constructed in such a way that the condition involves a mental representation of the agent. The representation cannot be one that just happens to represent the agent, because the rules must be constructed prior to any contingent knowledge of the world. For instance, the rules for intention formation cannot proceed in terms of descriptions that just happen to represent the agent, because it cannot be predicted ahead of time which of these will represent the agent. The rules themselves must require that the intentions involve a mental representation of the agent, and at the time the rules are being constructed it cannot be predicted what mental representations will represent the agent as a matter of contingent fact, so the rules for intention formation must employ a mental representation that represents the agent necessarily. That is precisely what *de se* representations are.

It is illuminating to illustrate this by considering chess-playing programs. Sophisticated chess-playing programs learn and get better as they play. Existing programs do not involve any kind of *de se* representation, and that may seem to be a counterexample to the claims just made. But the reason such programs need not involve *de se* representation is that, in the appropriate sense, everything in their vocabulary describes the situation of the chess-playing computer, and so there is no distinction to be drawn between its situation and the general state of the universe. Contrast this with a computer running a more sophisticated kind of chess-playing program that learns not just by playing but also by witnessing games played by other computers. In this case the computer must be able to distinguish between its own game and games it is merely witnessing. Its own game must somehow be tagged *as* its own. In effect, this involves *de se* representation.

Our conclusion is that practical reasoning requires an agent to form beliefs about his *own* current situation and form intentions about what to do if his current situation is of a particular sort. It does no good to have general beliefs about the state of the universe if those beliefs cannot be related to the agent, and it is precisely the latter that cannot be done without *de se* beliefs. *De se* representations are required in order to make epistemic and practical cognition mesh properly. Thus *de se* beliefs are essential in the construction of a sophisticated cognitive/conative machine. We take it that that is why they play a central role in human thought.

It is somewhat illuminating to consider just what we have to do to equip Oscar with *de se* thought. It might seem that in order to do this we have to provide him with a Cartesian ego and a way of perceiving it. But it takes little reflection to see that that is wrong. Providing his language of thought with *de se* representations is just a matter of including a primitive singular term in his language of thought and then wiring him up in such a way that sensory input results in his having *de se* beliefs, and rational deliberation results in *de se* intentions. There is nothing mysterious about this. It is all a matter of programming (or wiring). In particular, we do not have to equip Oscar with a "ghost" in his machine.

8. Conclusions

Our first task in this chapter was to reflect on the methodology of epistemology and on the status of epistemic naturalism. Then, in order to discover why human rational cognition works as it does, we asked the more general question, "What is required to build a cognitive agent with humanlike cognitive capabilities?" We were led to the conclusion that there is no obvious way to build such an agent without incorporating many features of human cognition into its cognitive architecture. Apparently there is really no alternative to many of the rather general features of human cognition. In the next chapter we will turn more directly to human epistemic norms and try to formulate some of the important ones carefully. We believe that in many cases it would be possible to argue that all sophisticated cognitive agents must be equipped with similar norms.

7
DIRECT REALISM

1. Introduction

1.1 Defending Direct Realism

In chapters five and six, we proposed a naturalistic account of procedural epistemic justification and epistemic norms. That account is about the nature of the norms. What remains is to describe their content. What are the actual norms governing human epistemic competence? Most of the book has been concerned with eliminating general theories of those norms. We claim to have shown that epistemic norms must be internalist. We have also argued that the doxastic assumption is false, which implies that at least some epistemic norms must appeal to internal states other than beliefs. Thus we are led to some form of nondoxastic internalism.

One of the objections raised in chapter three to holistic coherence theories is that they cannot explain the difference between justified and justifiable beliefs. That distinction turns essentially on the reasons for which the beliefs are held. The failure of holistic coherence theories shows that reasons and reasoning play an essential role in justification. The kind of structure proposed by foundationalism seems to be essentially correct with regard to the role it assigns to reasoning.

Where foundationalism, and doxastic theories in general, go wrong, is that they cannot accommodate perceptual knowledge. Consider vision. It begins with a two-dimensional pattern of retinal stimulation. Our visual processing converts that into a percept, and then the agent forms beliefs about its surroundings. The beliefs are, in some manner, derived from and based on the percept. There are, in principle, two ways this could work. The simplest is that having the percept could itself give rise to beliefs about physical objects perceived. A more complex cognitive architecture would have the percept give rise instead to a belief describing the percept, and then beliefs about physical objects would be inferred from the beliefs about percepts. The latter is the account preferred by the traditional foundationalist, but we have seen that as an account of human cognition, it is wrong. Human beings do not normally form beliefs about their percepts. They move instead directly from the percepts to beliefs about physical objects perceived.

Because the traditional foundationalist was enamored of the doxastic assumption, he thought it impossible to move, with epistemic justification, directly from percepts to beliefs about physical objects. In other words, he thought such epistemic norms were impossible. But as we have seen, that turned upon a deep misunderstanding of the way epistemic norms guide cognition. Once the intellectualist model is rejected, we can see that epistemic norms can in principle appeal to any internal states of the cognizer—not just to beliefs. Accordingly, there is no reason we cannot have epistemic norms licensing a cognitive move directly from percepts to beliefs about physical objects. The resulting theory looks much like traditional foundationalism, differing only in that the foundations consist of percepts rather than beliefs about percepts. A theory of this sort is what is called "direct realism". It must be acknowledged that the terminology is not very perspicuous, but what it is intended to capture is the idea that the move from percepts to beliefs about the real world is direct and unmediated by beliefs about the agent's internal states.[163]

The rejection of foundationalism is only a rejection of it as a description of human rational cognition. Humans do not work that way, but there is no apparent reason why there could not be cognitive agents with a somewhat different cognitive architecture that conforms to traditional foundationalism. Presumably, the reason humans are constructed so that they move directly from percepts to beliefs about the physical world is that it is usually unnecessary to form beliefs about percepts. We can form them when they are useful, but they are not usually useful, and having to form them in all cases would expend our limited cognitive resources unnecessarily.

1.2 Levels of Epistemological Theorizing

Epistemology is driven by attempts to answer the question, "How do you know?" In the last chapter we pointed out that this gives rise to investigations on several different levels. At the lowest level, philosophers investigate particular kinds of knowledge claims. At an intermediate level, topics are investigated that pertain to all or most of the specific kinds of knowledge discussed at the lowest level. At the highest level we find general epistemological theories that attempt to explain how knowledge in general is possible. One can be doing epistemology by working at any of these levels. The levels cannot be isolated from each other, however. Work at any level tends to presuppose something about the other levels. For example, work on inductive reasoning at least presupposes that reasoning plays a role in the acquisition or justification of beliefs and normally presupposes something about the structure of defeasible reasoning.

Reflection on high-level epistemological theories has typically proceeded in a rather abstract fashion. Defenders of theories like coherentism or probabilism have formulated their theories in very general terms, and have usually made only half-hearted attempts to show how they can accommodate the specific kinds of epistemic cognition required for knowl-

---

163. A more perspicuous term might be "nondoxastic foundationalism". Our main reason for resisting this terminology is to keep our taxonomy of epistemological theories simple. On our taxonomy, foundationalist theories fall within the broader category of doxastic theories, and so nondoxastic foundationalism is impossible.

edge of concrete subject matters. This can be regarded as a kind of *top-down epistemological theorizing*. Concentration on low-level theories is a kind of *bottom-up theorizing*. Neither top-down nor bottom-up theorizing can be satisfactory by itself. A necessary condition for the correctness of a low-level theory (e.g., a theory of inductive reasoning, or a theory of inference from perception) is that it must fit into a correct high-level theory. Focusing on the low-level theory by itself, without reference to a high-level theory into which it must fit, is theorizing in a relative vacuum and imposes too few constraints. Conversely, it is equally a necessary condition for the correctness of a high-level theory that it be possible to fill it out with low-level theories of specific kinds of epistemic cognition, and the only way to verify that this can be done is to do it. To be ultimately satisfactory, epistemological theorizing must combine top-down and bottom-up theorizing.

It seems likely that little progress can be made on low-level theories without presupposing something about high-level theories. Accordingly, the natural way to proceed in epistemology is to begin by giving general arguments for a high-level theory, and then to fill it out by constructing low-level theories compatible with it. Difficulties in constructing the low-level theories should lead to modification of the high-level theory, or in extreme cases, to its abandonment.

1.3 Filling Out Direct Realism

The development of the OSCAR Project follows the course we are proposing. The discussion in this book to this point provides the arguments for a high-level theory—direct realism. We have surveyed the different possible high-level structural theories, and raised what we regard as compelling objections to all but direct realism. Our conclusion is that direct realism, as the only survivor, must be correct. Accordingly, we take direct realism as the jumping-off point for the construction of low-level epistemological theories. Being able to construct low-level theories consonant with direct realism is a necessary condition for direct realism to be correct. Should it prove impossible to construct the low-level theories, direct realism would have to be abandoned. For this reason, the OSCAR Project takes its principal task to be that of constructing and implementing such low-level theories. That is, we are attempting to specify the low-level details of our epistemic norms with enough care so that a computer program can encode those norms. The end result will be a functioning artificial intellect—an artilect. The construction of such an artilect is a challenge that must ultimately be met by any satisfactory epistemological theory.[164] We take this opportunity to throw down the gauntlet to other

---

164. One of us (Cruz) is persuaded that there might be another way to work out the low-level details of an epistemological theory. He thinks that detailed psychological explanations will ultimately reveal our epistemic norms. On his view, psychological explanations are the result of a competence theory of human rationality expressed by the cognitive science researcher who is seeking to explain empirical results. In light of

epistemologists who disagree with us about direct realism. If they are to defend their theories adequately, coherentists, probabilists, etc., must not only answer the theoretical objections that we have leveled against them—they must show that it is possible to build an artilect founded on their theories of epistemic justification. We seriously doubt that this challenge can be met. On the other hand, direct realism is well on its way to meeting this challenge.

The task of constructing low-level theories is that of describing the various species of reasoning that can lead to justified beliefs about different subject matters. However, such a low-level theory will typically presuppose a mid-level theory about general aspects of deductive and defeasible reasoning. In particular, low-level theories must be implemented on top of an *inference engine*. Most of the work on the OSCAR Project over the last decade has concerned the formulation and implementation of a general theory of defeasible and deductive reasoning. The next section will describe that theory, and then the following sections will apply it in developing accounts of perceptual knowledge and certain kinds of temporal and causal reasoning. This constitutes a partial filling out of direct realism.[165]

We should offer one last comment by way of introducing this chapter. Philosophers are often criticized for shying away from offering detailed, empirically testable claims about their favorite subject matter. Somehow philosophers have internalized this criticism so well that they are skeptical of *one another* when they see their cohorts expending effort to say with precision what a theory commits them to. Perhaps philosophers think it is dangerous to propose anything detailed because it might be *clearly* refuted by future research, rendering their efforts obsolete. We think this is a wholly lamentable shift away from a kind of ambitious philosophy that takes itself to be deeply committed to understanding ourselves and the world. Should some or all of the low-level details of direct realism that we propose in this chapter be overturned, we welcome the opportunity to see where we went wrong.

2. Reasoning

2.1 *Reasons*

According to direct realism, epistemic norms must be able to appeal directly to our being in perceptual states and need not appeal to our having beliefs to that effect. In other words, there can be "half-doxastic"

---

this, he proposes the Quinean thesis that psychology should replace epistemology, though the replacement will contain a normative component in a way similar to the way that normativity is treated in this book. See Cruz's Ph.D. dissertation (1999).

165. Many details are omitted from the account given here. The details can be found in Pollock (1998).

epistemic connections between beliefs and nondoxastic states that are analogous to the "fully doxastic" connections between beliefs and beliefs that we call "reasons". We propose to call the half-doxastic connections "reasons" as well, but it must be acknowledged that this is stretching our ordinary use of the term 'reason'. The motivation for this terminology is that the logical structure of such connections is completely analogous to the logical structure of ordinary defeasible reasons. That is, the half-doxastic connections convey justification defeasibly, and the defeaters operate like the defeaters proposed by the foundations theory formulated in chapter two.

The treatment of reasons adopted in chapter two followed standard philosophical practice and took reasons to be the propositions believed rather than the states of believing those propositions. As long as we were concerned exclusively with reasoning from beliefs, that was unassailable. But once reasoning from nondoxastic states is allowed, the *reason-for* relation must be described differently because different kinds of states can have the same propositional content but support different inferences. For example, the belief that there is something red before me, the percept of there being something red before me, and the desire that there be something red before me, all have the propositional content "There is something red before me", but being in those states licenses different inferences. To accommodate this, we modify our earlier definitions of 'reason' and 'defeater' as follows:

DEFINITION:
A state M of a person S is a *reason* for S to believe Q if and only if it is logically possible for S to become justified in believing Q by believing it on the basis of being in the state M.

DEFINITION:
If M is a reason for S to believe Q, a state M^* is a *defeater* for this reason if and only if the combined state consisting of being in both the state M and the state M^* at the same time is not a reason for S to believe Q.

Reasons are always reasons *for* beliefs, but the reasons themselves need not be beliefs. For instance, direct realism can handle the problem of perception by adopting nondoxastic reasons such as the following:

x's looking red to S is a reason for S to believe that x is red.

This means that the perceptual state itself is the reason, and not a belief about the perceptual state. In the most common case, the state M will consist of believing some proposition P, in which case we can speak loosely of P itself being a defeasible reason for Q.

2.2 Defeaters

Defeasible reasons are those for which there are (possible) defeaters. There are two kinds of defeaters for defeasible reasons. The simplest is a reason for denying the conclusion:

DEFINITION:
If M is a defeasible reason for S to believe Q, M^* is a *rebutting defeater* for this reason if and only if M^* is a defeater (for M as a reason for S to believe Q) and M^* is a reason for S to believe $\sim Q$.

For instance, a counterexample to an inductive generalization is a rebutting defeater for the inductive evidence.

It has often been overlooked that there is a second kind of defeater for defeasible reasons. Such defeaters attack the connection between the reason and the conclusion rather than attacking the conclusion itself. In chapter one we gave the example of a pollster attempting to predict what proportion of residents of Indianapolis will vote for the Republican gubernatorial candidate in the next election. She randomly selects a sample of voters and determines that 87 percent of those polled intend to vote Republican. This gives her a defeasible reason for thinking that approximately 87 percent of all Indianapolis voters will vote Republican. But then it is discovered that purely by chance, her randomly chosen sample turned out to consist exclusively of voters with incomes over $100,000 per year. This constitutes a defeater for the inductive reasoning, but it is not a reason for thinking it false that approximately 87 percent of the voters will vote Republican. The discovery is neutral to that question. Instead, it is a reason for doubting or denying that we would not have the inductive evidence unless the conclusion were true. The defeater attacks the connection between the evidence and the conclusion rather than attacking the conclusion itself. More generally:

DEFINITION:
If believing P is a defeasible reason for S to believe Q, M^* is an *undercutting defeater* for this reason if and only if M^* is a defeater (for believing P as a reason for S to believe Q) and M^* is a reason for S to doubt or deny that P would not be true unless Q were true.[166]

When the reason is a nondoxastic state, we must define undercutting defeat slightly differently:

---

166. The distinction between rebutting defeaters and undercutting defeaters originates in Pollock (1970), and was explored further in Pollock (1974) where they were called "Type I" and "Type II" defeaters, respectively.

DEFINITION:
If M is a nondoxastic state that is a defeasible reason for S to believe Q, M^* is an *undercutting defeater* for this reason if and only if M^* is a defeater (for M as a reason for S to believe Q) and M^* is a reason for S to doubt or deny that he or she would not be in state M unless Q were true.

It will be convenient to abbreviate "It is false that P would not be true unless Q were true" as "$(P \otimes Q)$". This can be read more simply as "P does not guarantee Q". Undercutting defeaters are reasons for $(P \otimes Q)$.

2.3 Justified Beliefs

In direct realism, beliefs get justified by reasoning. Reasoning proceeds by stringing reasons together into arguments. Modifying the account given in chapter two, we can think of an argument as a finite sequence of propositions (representing beliefs) and nondoxastic mental states ordered in such a way that each member of the sequence is either (1) a nondoxastic mental state, or (2) such that there is a proposition (or set of propositions) or a nondoxastic state earlier in the sequence that is a reason for P.[167] A person *instantiates* an argument if and only if he is in the nondoxastic states and he believes each proposition in the argument on the basis of reasons for it that occur earlier in the argument. Recalling our earlier discussion, an argument *supports* a proposition if and only if that proposition is the final proposition in the argument.

This is the point in the discussion of justification where we have to introduce defeaters. It seems initially reasonable to suppose that belief in P is justified for a person S if and only if S instantiates an argument supporting P, but this proposal fails because it overlooks defeasibility. To illustrate, suppose a person S simultaneously instantiates each of the arguments shown in figure 7.1, where P_1 and Q_1 are nondoxastic states. The conclusion of the second argument is an undercutting defeater for the second step of the first argument. Under the circumstances, it seems that S is not justified in believing P_3. Of course, that could change if S also instantiates an argument supporting a defeater for some step of the second argument. That would reinstate the first argument.

To handle defeasibility in a general way, we must recognize that arguments can defeat one another and that a defeated argument can be reinstated if the arguments defeating it are defeated in turn. We can think of arguments as provisional vehicles of justification, and then give rules for when these provisional vehicles succeed in conferring justification

---

167. This is probably simplistic in at least one respect. It seems likely that we should allow arguments to contain subsidiary arguments. For example, an argument might contain a subsidiary argument supporting a conditional by conditional proof, or a subsidiary argument supporting a negation by *reductio ad absurdum*. The additional sophistications that this requires are not relevant to the present discussion, so we will ignore them for now.

and when they do not. The way in which this works is complicated, and somewhat controversial. A complete theory of justified belief must take account of the fact that reasons can differ in strength, some supplying more support for their conclusions than others. But that adds significant complexity to the theory of justified belief, so to keep things simple, we will make for a moment the simplifying assumption that all reasons are of the same strength.[168]

Figure 7.1 Defeat of one argument by another

Reasoning proceeds by adding steps to arguments, constructing longer arguments out of shorter arguments. The process starts with one-line arguments consisting of single percepts. The shorter arguments from which a longer argument is constructed are its *subsidiary arguments*. Let us say that an *inference-graph* is a set G of arguments such that for any argument in G, all of its subsidiary arguments are in G as well. Drawing G as a graph, we will label the nodes with the conclusions of the arguments. With the exception of one-line arguments, an argument is always constructed by making an inference which adds a single step to some group of subsidiary arguments. The latter form the *basis* of the inference, and will be called *the basis of* the new argument. For example, if we reason as in figure 7.1 from P_1 to P_2 to P_3, the one-step argument asserting P_1 is the basis of the argument from P_1 to P_2, and the argument from P_1 to P_2 is the basis of the argument from P_1 to P_2 to P_3. Arrows are drawn between the nodes of the inference graph to represent inference. This allows us to read off the structure of the argument, as in figure 7.1. A *defeater for an argument* is an argument supporting a defeater for its final step. Intuitively, the way an argument gets defeated is by either (1) being based on a defeated subsidiary argument, or (2) having an undefeated defeater. So

---

168. For the complete theory, taking account of variable reason strengths, see Pollock (1995) and below. For a comparison with other theories of defeasible reasoning, see Prakken and Vreeswijk (2000).

we define a *status-assignment* for an inference-graph to be an assignment of "defeated" and "undefeated" to the arguments in the inference-graph in accordance with this rule. More precisely, we define:

A *partial-status-assignment* for an inference-graph G is an assignment of "defeated" and "undefeated" to a subset of the arguments in G such that for each argument A in G:
1. if A is a one-line argument (i.e., a single percept), A is assigned "undefeated";
2. if some defeater for A is assigned "undefeated", or some member of the basis of A is assigned "defeated", A is assigned "defeated";
3. if all defeaters for A are assigned "defeated" and all members of the basis of A are assigned "undefeated", A is assigned "undefeated".

Partial-status-assignments need not assign defeat statuses to every member of G. So let us define:

A *status-assignment* for an inference-graph G is a maximal partial-status-assignment, i.e., a partial-status-assignment not properly contained in any other partial-status-assignment.

Then we define:

An argument A is *undefeated relative to* an inference-graph G of which it is a member if and only if every status-assignment for G assigns "undefeated" to A.

A belief is justified if and only if it is supported by an argument that is undefeated relative to the inference-graph that represents the believer's current epistemological state.

To see how this analysis works, consider some simple examples. Suppose first that a person reasons as in figure 7.1, and does not perform any other relevant reasoning. Then there is just one status-assignment. It assigns "undefeated" to the arguments to P_1, P_2, Q_1, Q_2, and $(P_2 \otimes P_3)$. Because the latter is assigned "undefeated", the argument to P_3 is assigned "defeated".

But now consider a more complicated example. Let us assume that someone's telling us something gives us a reason for believing it. Suppose we have percepts P_1 leading us to believe that Jones says it is raining, and P_2 leading us to believe that Smith says it is not raining. Then we have equally good reasons for believing both that it is and that it isn't raining. What should we believe? Intuitively, we should withhold belief in either conclusion until we get more evidence. The relevant reasoning can be diagrammed as in figure 7.2, where the fuzzy arrow symbolizes defeat of one argument by another. There are two status-assignments for this inference-graph. One assigns "undefeated" to P_1, P_2, "Jones says that it is raining", "Smith says that it is not raining", and "It is raining",

and assigns "defeated" to "It is not raining". The other assigns "undefeated" to P_1, P_2, "Jones says that it is raining", "Smith says that it is not
raining", and "It is not raining", and assigns "defeated" to "It is raining".
Accordingly, the arguments for "It is raining" and "It is not raining" are
both assigned "defeated" by some status-assignment, and so both arguments are defeated relative to the inference-graph. This captures our
intuition that we should not accept either conclusion. A case like this, in
which two or more arguments defeat each other, is called "collective
defeat".

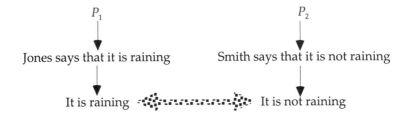

Figure 7.2 Collective defeat

Finally, consider a still more complicated example, diagrammed by
figure 7.3. Suppose Smith and Jones each accuse the other of lying.
Then we have a case of collective defeat like figure 7.2. We should
remain agnostic about whether either is a liar. Under these circumstances,
if Smith also tells us that it is raining, it would be unreasonable to take
his word for it. This illustrates that the conclusion that Smith is a liar
can defeat the inference from Smith's saying that it is raining to its
raining, even though the conclusion that Smith is a liar is defeated. What
is crucial to this example is that the defeat is a case of collective defeat.
There are two status-assignments for this inference-graph. One assigns
"defeated" to "Smith is a liar" and "undefeated" to "Jones is a liar", and
the other does the reverse. But notice that the assignment assigning
"undefeated" to "Smith is a liar" assigns "defeated" to "It is raining".
Thus there is an assignment in which "It is raining" is assigned "defeated",
and so that conclusion is not justified. This shows that an argument can
be defeated because there is a status-assignment assigning "defeated" to
it, but it can retain the ability to defeat other arguments if there is also a
status-assignment assigning "undefeated" to it. Such arguments are said
to be *provisionally defeated*. An argument is *defeated outright* if every status
assignment assigns "defeated" to it. Arguments that are defeated outright
do not have the power to defeat other arguments.

This account of defeasible reasoning provides the inferential machinery
upon which to build low-level theories of the epistemic norms governing
specific kinds of knowledge. It is the middle-level between our abstract
defense of direct realism and the low-level reasoning schemas. We will
illustrate the low-level theory with discussions of perception, certain
kinds of temporal and causal reasoning, and some varieties of probabilistic
and inductive reasoning.

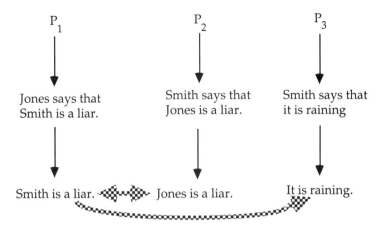

Figure 7.3 Provisional defeat

3. Perception

Recall that the problem of perception is the problem of explaining how we can gain knowledge of the external world through perception. In earlier chapters, we discussed and rejected all of the classical attempts to solve the problem of perception. Their failure is the principal motivation for direct realism. The fundamental principle underlying direct realism is that perception provides reasons for judgments about the world, and the inference is made directly from the percept rather than being mediated by a (basic) belief about the percept. This principle can be formulated very simply as the following reason-schema:

PERCEPTION
Having a percept at time t with the content P is a defeasible reason for the cognizer to believe P-at-t.

Traditional philosophers were worried about explaining what justifies an inference like this. The answer forthcoming from the account of human rationality proposed above is that nothing justifies this. This is one of the basic principles of rational cognition that make up the human cognitive architecture. This is partially constitutive of our knowledge of how to cognize. It cannot be derived from anything more basic. This principle, or something like it, must be present in the cognitive architecture of any agent that is capable of reacting to the way the world is in a sophisticated manner. By definition, rational agents direct their activity in response to their beliefs about the way the world is, so some such principle is an essential ingredient of the rational architecture of any rational agent.

When giving a low-level account of a species of defeasible reasoning, it is as important to characterize the defeaters for the defeasible reasons as it is to state the reasons themselves. The only obvious undercutting defeater for **PERCEPTION** is a reliability defeater, which is of a general sort applicable to all defeasible reasons. Reliability defeaters result from observing that the inference from P to Q is not, under the present circumstances, reliable. Precisely:

PERCEPTUAL-RELIABILITY
Where R is projectible, "R-at-t, and the probability is low of P's being true given R and that I have a percept with content P" is an undercutting defeater for **PERCEPTION**.

The projectibility constraint in this principle is a perplexing one. To illustrate its need, suppose I have a percept of a red object, and am in improbable but irrelevant circumstances of some type C_1. For instance, C_1 might consist of my having been born in the first second of the first minute of the first hour of the first year of the twentieth century. Let C_2 be circumstances consisting of wearing rose-colored glasses. When I am wearing rose-colored glasses, the probability is not particularly high that an object is red just because it looks red, so if I were in circumstances of type C_2, that would quite properly be a reliability defeater for a judgment that there is a red object before me. However, if I am in circumstances of type C_1 but not of C_2, there should be no reliability defeater. The difficulty is that if I am in circumstances of type C_1, then I am also in the disjunctive circumstances $(C_1 \vee C_2)$. Furthermore, the probability of being in circumstances of type C_2 given that one is in circumstances of type $(C_1 \vee C_2)$ is very high, so the probability is not high that an object is red given that it looks red to me but I am in circumstances $(C_1 \vee C_2)$. Consequently, if $(C_1 \vee C_2)$ were allowed as an instance of R in **PERCEPTUAL-RELIABILITY**, being in circumstances of type C_1 would suffice to indirectly defeat the perceptual judgment.

The preceding example shows that the set of circumstance-types appropriate for use in **PERCEPTUAL-RELIABILITY** is not closed under disjunction. This is a general characteristic of projectibility constraints. The need for a projectibility constraint in induction was discussed earlier in this book. It was shown in Pollock (1990) that the same constraint occurs throughout probabilistic reasoning, and the constraint on induction can be regarded as derivative from a constraint on the statistical syllogism.[169] However, similar constraints occur in other contexts and do not appear to be derivative from the constraints on the statistical syllogism. The constraint on reliability defeaters is one example of this, and several other examples will be given below when we turn to other epistemic norms.

---

169. The material on projectibility in Pollock (1990) has been collected into a paper and reprinted in Pollock (1994).

4. Implementation

With the middle-level account of reasoning and the reason-schema for perception before us, we are now in a position to share some of the implementational details. This section can be skipped without loss of understanding. It is included for those who want to see how reason-schemas are actually implemented in OSCAR.[170]

Reasoning in OSCAR consists of the construction of natural-deduction-style arguments, using both deductive inference rules and defeasible reason-schemas. Premises are input to the reasoner (either as background knowledge or as new percepts), and queries are passed to the reasoner. OSCAR performs bidirectional reasoning. The reasoner reasons forward from the premises and backward from the queries. The queries are "epistemic interests", and backward reasoning can be viewed as deriving interests from interests. Conclusions are stored as nodes in the inference-graph (*inference-nodes*).

Reasoning proceeds in terms of reasons. *Backward-reasons* are used in reasoning backward, and *forward-reasons* are used in reasoning forward. Forward-reasons are data-structures with the following fields:

- reason-name
- forward-premises — a list of forward-premises
- backward-premises — a list of backward-premises
- reason-conclusion — a formula
- defeasible-rule — T if the reason is a defeasible reason, NIL otherwise[171]
- reason-variables — variables used in pattern-matching to find instances of the reason-premises
- reason-strength — a real number between 0 and 1, or an expression containing some of the reason-variables and evaluating to a number
- reason-description — an optional string describing the reason

Forward-premises are data-structures encoding the following information:

- fp-formula — a formula
- fp-kind — :inference, :percept, or :desire (the default is :inference)
- fp-condition — an optional constraint that must be satisfied by an inference-node for it to instantiate this premise
- clue? — explained below (in section six)

Similarly, *backward-premises* are data-structures encoding the following information:

- bp-formula

---

170. The code for the OSCAR defeasible reasoner, and the code for the reasoning-schemas discussed here, can be downloaded from http://www.u.arizona.edu/~pollock/.

171. NIL is "the false", sometimes written "⊥".

- bp-kind
- bp-condition

The use of the premise-kind is to check whether the formula from which a forward inference proceeds represents a desire, percept, or the result of an inference.

Backward-reasons will be data-structures encoding the following information:

- reason-name
- forward-premises
- backward-premises
- reason-conclusion — a formula
- reason-variables — variables used in pattern-matching to find instances of the reason-premises
- strength — a real number between 0 and 1, or an expression containing some of the reason-variables and evaluating to a number
- defeasible-rule — T if the reason is a defeasible reason, NIL otherwise
- reason-condition — a condition that must be satisfied by the sequent of interest before the reason is to be deployed

Simple forward-reasons have no backward-premises, and *simple backward-reasons* have no forward-premises. Given inference-nodes that instantiate the premises of a simple forward-reason, the reasoner infers the corresponding instances of the conclusions. Similarly, given an interest that instantiates the conclusion of a simple backward-reason, the reasoner adopts interest in the corresponding instances of the backward-premises. Given inference-nodes that discharge those interests, an inference is made to the conclusions from those inference-nodes.

In deductive reasoning, with the exception of a rule of reductio ad absurdum, we are unlikely to encounter any but simple forward- and backward-reasons.[172] However, the use of backward-premises in forward-reasons and the use of forward-premises in backward-reasons provides an invaluable form of control over the way reasoning progresses. This will be illustrated at length below. *Mixed* forward- and backward-reasons are those having both forward- and backward-premises. Given inference-nodes that instantiate the forward-premises of a mixed forward-reason, the reasoner does not immediately infer the conclusion. Instead the reasoner adopts interest in the corresponding instances of the backward-premises, and an inference is made only when those interests are discharged. Similarly, given an interest instantiating the first conclusion of a mixed backward-reason, interests are not immediately adopted in the backward-premises. Interests in the backward-premises are adopted only when inference-nodes are constructed that instantiate the forward-premises.

There can also be *degenerate backward-reasons* that have only forward-

---

172. This is discussed at greater length in chapter two of Pollock (1995).

premises. In a degenerate backward-reason, given an interest instantiating the first conclusion, the reasoner then becomes "sensitive to" inference-nodes instantiating the forward-premises, but does not adopt interest in them (and thereby actively search for arguments to establish them). If appropriate inference-nodes are produced by other reasoning, then an inference is made to the conclusions. Degenerate backward-reasons are thus much like simple forward-reasons, except that the conclusion is only drawn if there is an interest in it.

Reasons are most easily defined in OSCAR using the macros DEF-FORWARD-REASON and DEF-BACKWARD-REASON:

(def-forward-reason *symbol*
    :forward-premises *list of formulas optionally interspersed with expressions of the form (:kind ...) or (:condition ...)*
    :backward-premises *list of formulas optionally interspersed with expressions of the form (:kind ...) or (:condition ...)*
    :conclusion *a formula*
    :strength *number or a an expression containing some of the reason-variables and evaluating to a number.*
    :variables *list of symbols*
    :defeasible? *T or NIL (NIL is the default)*)

(def-backward-reason *symbol*
    :conclusion *a formula*
    :forward-premises *list of formulas optionally interspersed with expressions of the form (:kind ...) or (:condition ...)*
    :backward-premises *list of formulas optionally interspersed with expressions of the form (:kind ...) or (:condition ...)*
    :condition *this is a predicate applied to the binding produced by the target sequent*
    :strength *number or an expression containing some of the reason-variables and evaluating to a number.*
    :variables *list of symbols*
    :defeasible? *T or NIL (NIL is the default)*)

The use of these macros will be illustrated below.

Epistemic reasoning begins from contingent information input into the system in the form of percepts. Percepts are encoded as structures with the following fields:

- percept-content—a formula, with temporal reference built in.
- percept-strength—a number between 0 and 1, indicating how strong a reason the percept provides for the conclusion of a perceptual inference.
- percept-date—a number.

When a new percept is presented to OSCAR, an inference-node of kind :percept is constructed, having a node-formula that is the percept-content of the percept (this includes the percept-date). This inference-node is then inserted into the inference-queue for processing.

Using the tools described above, we can implement PERCEPTION as
a simple forward-reason:

```
(def-forward-reason PERCEPTION
  :forward-premises "(p at time)" (:kind :percept)
  :conclusion "(p at time)"
  :variables p time
  :defeasible? t
  :strength .98)
```

The strength of .98 has been chosen arbitrarily.
To proceed further, we must say something about the way reason-
strengths are measured in OSCAR. Some reasons are better than others.
In OSCAR, reason-strengths range from 0 to 1. Reason-strengths are
calibrated by comparing them with the *statistical syllogism* (discussed
further in section eight) according to which, when $r > 0.5$, "Bc & prob(A/B)
= r" is a defeasible reason for "Ac", the strength of the reason being a
function of r.[173] A reason of strength r is taken to have the same strength
as an instance of the statistical syllogism from a probability of $2 \cdot (r - 0.5)$.
This maps reason-strengths in the interval $[0,1]$ to probabilities in the
interval $[0.5,1]$. The inference rule PERCEPTION will have some strength
r, and for artificial agents this may vary from agent to agent. The value
of r should correspond roughly to the reliability of an agent's system of
perceptual input in the circumstances in which it normally functions.
Reformulating PERCEPTUAL-RELIABILITY to take account of reason-
strengths, we get:

PERCEPTUAL-RELIABILITY
Where R is projectible, r is the strength of PERCEPTION, and $s <$
$0.5 \cdot (r + 1)$, "R-at-t, and the probability is less than or equal to s of
P's being true given R and that I have a percept with content P"
is an undercutting defeater for PERCEPTION.

It seems clear that this should be treated as a backward-reason. That is,
given an interest in the undercutting defeater for PERCEPTION, this reason-
schema should be activated, but if the reasoner is not interested in the
undercutting defeater, this reason-schema should have no effect on the
reasoner. However, treating this as a simple backward-reason is impossible, because there are no constraints (other than projectibility) on R. We
do not want interest in the undercutting defeater to lead to interest in
every projectible R. Nor do we want the reasoner to spend its time
trying to determine the reliability of perception given everything it happens to know about the situation. This can be avoided by making this a
degenerate backward-reason (no backward-premises), taking R-at-t

---

173. This is a slight oversimplification. See section eight for a more detailed
discussion of the statistical syllogism.

(where *t* is the time of the percept) and the probability premise to be a forward-premise. This suggests the following definition:

```
(def-backward-undercutter PERCEPTUAL-RELIABILITY
  :defeatee PERCEPTION
  :forward-premises
  "((the probability of p given ((I have a percept with content p) & R)) ≤ s)"
   (:condition  (and (s < 0.99) (projectible R)))
   "(R at time)"
  :variables p time R s
  :defeasible? t)
```

(DEF-BACKWARD-UNDERCUTTER is a variant of DEF-BACKWARD-REASON that computes the reason-conclusions for us.)

A problem remains for this implementation. PERCEPTUAL-RELIABILITY requires us to know that *R* is true at the time of the percept. We will typically know this only by inferring it from the fact that *R* was true earlier. The nature of this inference is the topic of the next section. Without this inference, it is not possible to give interesting illustrations of the implementation just described, so that will be postponed until section six.

5. Temporal Projection

The reason-schema PERCEPTION enables a cognizer to draw conclusions about one's current surroundings on the basis of one's current percepts. It is natural to suppose that this suffices to provide a person with the basic data needed to reason about the world, and if we supplement this with reason-schemas enabling one to reason inductively, together perhaps with some special purpose reason-schemas pertaining to reasoning about particular subject matters like other minds, then the cognizer will be able to reason her way to a rich model of the world. However, there is a gap in this reasoning that has been overlooked by epistemologists. The problem is that perception is really a form of sampling. It is not possible to continually monitor the entire state of the world perceptually. All we can do is sample small space-time chunks of the world and then make inferences from combinations of these samplings. There is a surprising difficulty connected with making inferences from combinations of perceptual samplings. This can be illustrated by considering a robot whose task is to visually check the readings of two meters and then press one of two buttons depending upon which reading is higher. This should not be a hard task, but if we assume that the robot can only look at one meter at a time, it will not be able to acquire the requisite information about the meters using only the reason-schema PERCEPTION. The robot can look at one meter and draw a conclusion about its value, but when the robot turns to read the other meter, it no longer has a percept of the first and so is no longer in a position to hold a justified belief about what that meter reads *now*. The cognitive architecture of the robot must be supplemented with some reason for believing that the first meter still reads what it read a moment ago. In other words, the robot must have some basis for regarding the meter reading as a *stable property*—one that tends not to change quickly over time.

One might suppose that a cognizer that can reason inductively would be able to discover that properties like meter readings are stable over at least short intervals. However, it turns out to be impossible to perform the requisite inductive reasoning without already presupposing the stability at issue. To say that a property is stable is to say that objects possessing it tend to retain it. To confirm this inductively, a cognizer would have to re-examine the same object at different times and determine whether the property has changed. The difficulty is that in order to do this, the cognizer must be able to reidentify the object as the same object at different times. Although this is a complex matter, it seems clear that a cognizer must make essential use of the perceptible properties of objects in reidentifying them. If the perceptible properties of objects fluctuated rapidly and unpredictably, it would be impossible to reidentify them. The upshot of this is that it is epistemically impossible to investigate the stability of perceptible properties inductively without presupposing that most of them tend to be stable.[174] Some such assumption of stability must be built into the cognitive architecture of an agent capable of learning about the world perceptually. On the other hand, given an assumption of stability, the agent can use induction to refine it by discovering that some perceptible properties are more stable than others, that particular properties tend to be unstable under specifiable circumstances, etc. Apparently, a rational agent must come equipped with reason-schemas of the following sort for at least some choices of P:

(7.1)   If $t_0 < t_1$, believing P-at-t_0 is a defeasible reason for the agent to believe P-at-t_1.

A stable property is one such that if it holds at one time, the probability is high that it will continue to hold at a later time. Let ρ be the probability that P will hold at time $t+1$ given that it holds at time t. Assuming independence, it follows that the probability that P will hold at time

---

174. A more detailed presentation of this argument can be found in chapter six of Pollock (1974).

$(t+\Delta t)$ given that it holds at time t is $\rho^{\Delta t}$. In other words, the strength of the presumption that a stable property will continue to hold over time decays as the time interval increases. This has very important consequences for perception. Consider *the perceptual updating problem* in which an agent has a percept of P at time t_0, and a percept of $\sim P$ at a later time t_1. What an agent *should* conclude (defeasibly) under these circumstances is that the world has changed between t_0 and t_1, and although P was true at t_0, it is no longer true at t_1 and hence no longer true at a later time t_2. If we attempt to reconstruct this reasoning using principle (7.1) without taking account of decaying reason strengths, we get the wrong answer. We get the reasoning diagrammed in figure 7.4, where the dashed arrows represent defeasible inferences. The conclusion P-at-t_2 and the conclusion $\sim P$-at-t_2 are both inferred defeasibly, but they contradict each other, so each constitutes a rebutting defeater for the other. This becomes a case of collective defeat, with both conclusions being defeated. This would make perceptual updating impossible. However, once we take account of reason strengths, the problem evaporates. To reflect the fact that the probability of $\rho^{\Delta t}$ decreases as Δt increases, we can build into the reason-schema the stipulation that the reason-strength decays as Δt increases:

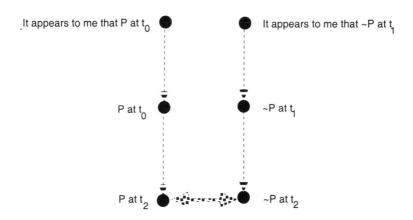

It appears to me that P at t_0 It appears to me that ~P at t_1

P at t_0 ~P at t_1

P at t_2 ~P at t_2

Figure 7.4 The perceptual updating problem

(7.2)   Believing P-at-t is a defeasible reason for the agent to believe P-at-$(t+\Delta t)$, the strength of the reason being a monotonic decreasing function of Δt.

Principle (7.2) has the consequence that the support for P-at-t_2 is weaker than the support for $\sim P$-at-t_2, and hence the former is defeated and it is defeasibly reasonable to accept the latter.

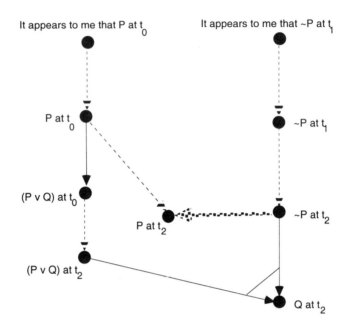

Figure 7.5 The need for a temporal projectibility constraint.

Although principle (7.2) handles the example of figure 7.4 correctly, it is not yet an adequate formulation of a reason-schema for temporal projection. The difficulty is that it is subject to a projectibility problem, much like that discussed above in connection with PERCEPTUAL-RELIABILITY. This is illustrated by the example diagrammed in figure 7.5, where solid arrows symbolize deductive inferences, and bars connecting arrows indicate that the inference is from multiple premises. Let P and Q be unrelated propositions. Suppose we know that P is true at t_0, and false at the later time t_1. Consider a third time t_2 later than t_1. P-at-t_0 gives us a defeasible reason for expecting P-at-t_2, but ~P-at-t_1 gives us a stronger reason for expecting ~P-at-t_2, because $(t_2 - t_1) < (t_2 - t_0)$. Thus an inference to P-at-t_2 is defeated, but an inference to ~P-at-t_2 is undefeated. However, from P-at-t_0 we can deductively infer $(P \vee Q)$-at-t_0. Without any restrictions on the proposition-variable in temporal-projection, $(P \vee Q)$-at-t_0 gives us a defeasible reason for expecting $(P \vee Q)$-at-t_2. Given the inference to ~P-at-t_2, we can then infer Q-at-t_2. In this inference-graph, the conclusion Q-at-t_2 is undefeated. But this is unreasonable. Q-at-t_2 is inferred from $(P \vee Q)$-at-t_2. $(P \vee Q)$ is expected to be true at t_2 only because it was true at t_0, and it was only true at t_0 because P was true at t_0. This makes it reasonable to believe $(P \vee Q)$-at-t_2 only insofar as it is reasonable to believe P-at-t_2, but the latter is defeated. This example illustrates clearly that temporal-projection does not work equally well for all propositions. In particular, it does not work for disjunctions.

This appears to be a projectibility problem, analogous to that discussed above in connection with reliability defeaters. In temporal projection, the use of arbitrary disjunctions, and other non-projectible constructions, must be precluded. Using the concept of temporal-projectibility, temporal projection should be reformulated as follows:

TEMPORAL-PROJECTION
If P is temporally-projectible then believing P-at-t is a defeasible reason for the agent to believe P-at-$(t+\Delta t)$, the strength of the reason being a monotonic decreasing function of Δt.

It is unclear precisely what the connection is between the projectibility constraint involved in temporal projection and that involved in induction, so we have referred to it neutrally as "temporal-projectibility". Notice that in temporal-unprojectibility, disjunctions are not the only culprits. The ascriptions of properties to objects will generally be projectible, but the negations of such ascriptions need not be. For instance, "x is red" would seem to be temporally projectible. But "x is not red" is equivalent to a disjunction "x is blue or green or yellow or orange or ...", and as such it would seem to be temporally unprojectible. On the other hand, there are "bivalent" properties, like "dead" and "alive" for which the negation of an ascription is projectible because it is equivalent to ascribing the other (temporally-projectible) property. For a complete theory of this reasoning, we must supplement TEMPORAL-PROJECTION with an analysis of temporal-projectibility. Unfortunately, that proves no easier than providing an analysis of projectibility as it occurs in induction. We do not, at this time, have an analysis to propose.

TEMPORAL-PROJECTION must also be supplemented with an account of defeaters for this defeasible inference, but we will not pursue that here.[175]

6. Implementing Temporal Projection

This section is included for those interested in implementation. Others should skip to section seven.

In order to implement TEMPORAL-PROJECTION, we must have a test for the temporal-projectibility of formulas. Lacking a theory of temporal-projectibility, we will finesse this by assuming that atomic formulas, negations of atomic formulas whose predicates are on a list *bivalent-predicates*, and conjunctions of the above, are temporally projectible. This will almost certainly be inadequate in the long run, but it will suffice for testing most features of the proposed reason-schemas.

---

175. For more on this, see Pollock (1998).

It seems clear that TEMPORAL-PROJECTION must be treated as a backward-reason. That is, given some belief P-at-t, we do not want the reasoner to automatically infer P-at-$(t+\Delta t)$ for every one of the infinitely many times $\Delta t > 0$. An agent should only make such an inference when the conclusion is of interest. For the same reason, the premise P-at-t should be a forward-premise rather than a backward-premise—we do not want the reasoner adopting interest in P-at-$(t-\Delta t)$ for every $\Delta t > 0$. Making TEMPORAL-PROJECTION a backward reason has the effect that when the reasoner adopts interest in P-at-t, it will check to see whether it already has a conclusion of the form P-at-t_0 for $t_0 < t$, and if so it will infer P-at-t. This produces a degenerate backward-reason:

```
(def-backward-reason TEMPORAL-PROJECTION
  :conclusion "(p at time)"
  :condition (and (temporally-projectible p) (numberp time))
  :forward-premises
    "(p at time0)"
    (:condition (and (time0 < time*) ((time* - time0) < log(.5)/log(.99))))
  :variables p time0 time
  :defeasible? T
  :strength (- (* 2 (expt .99 (- time time0))) 1))
```

Here we have arbitrarily set $\rho = .99$.

As an illustration combing the implementation of reasoning with the reason-schemas for perception and temporal projection, consider the perceptual updating problem. This is an printout of OSCAR solving the problem:

```
================================================================
Inputs:
      (the color of Fred is red) : at cycle 1 with justification 1.0
      (the color of Fred is blue) : at cycle 30 with justification 1.0


Ultimate epistemic interests:
      (? x)((the color of Fred is x) at 50)    degree of interest = 0.5
================================================================
THE FOLLOWING IS THE REASONING INVOLVED IN THE SOLUTION
Nodes marked DEFEATED have that status at the end of the reasoning.


                 # 1
                 interest: ((the color of Fred is y0) at 50)
                 This is of ultimate interest
||||||||||||||||||||||||||||||||||||||||||||||||||||||||||||||||||||||||||||||||||||||||||||||||
It appears to me that ((the color of Fred is red) at 1)
||||||||||||||||||||||||||||||||||||||||||||||||||||||||||||||||||||||||||||||||||||||||||||||||
                 # 1
It appears to me that ((the color of Fred is red) at 1)
```

2
((the color of Fred is red) at 1)
Inferred by:
       support-link #1 from { 1 } by PERCEPTION
undefeated-degree-of-support = 0.98
3
((the color of Fred is red) at 50)   DEFEATED
undefeated-degree-of-support = 0.904
Inferred by:
       support-link #2 from { 2 } by TEMPORAL-PROJECTION
       defeaters: { 7 } DEFEATED
This discharges interest 1
               # 5
               interest: ~((the color of Fred is red) at 50)
               Of interest as a defeater for support-link 2 for node 3

       ===
       Justified belief in ((the color of Fred is red) at 50)
       with undefeated-degree-of-support 0.904
       answers #<Query 1: (? x)((the color of Fred is x) at 50)>
       ===

|||||||||||||||||||||||||||||||||||||||||||||||||||||||||||||||||||||||||||||||||||||||||||||||||||||||||||||
It appears to me that ((the color of Fred is blue) at 30)
|||||||||||||||||||||||||||||||||||||||||||||||||||||||||||||||||||||||||||||||||||||||||||||||||||||||||||||
4
It appears to me that ((the color of Fred is blue) at 30)
5
((the color of Fred is blue) at 30)
Inferred by:
       support-link #3 from { 4 } by PERCEPTION
undefeated-degree-of-support = 0.98
6
((the color of Fred is blue) at 50)
Inferred by:
       support-link #4 from { 5 } by TEMPORAL-PROJECTION defeaters: { 8 }
undefeated-degree-of-support = 0.960
This discharges interest 1
               # 9
               interest: ~((the color of Fred is blue) at 50)
               Of interest as a defeater for support-link 4 for node 6

       ===
       Justified belief in ((the color of Fred is blue) at 50)
       with undefeated-degree-of-support 0.960
       answers #<Query 1: (? x)((the color of Fred is x) at 50)>
       ===

7
~((the color of Fred is red) at 50)
Inferred by:

support-link #5 from { 6 } by INCOMPATIBLE-COLORS
undefeated-degree-of-support = 0.960
defeatees: { link 2 for node 3 }
vvvvvvvvvvvvvvvvvvvvvvvvvvvvvvvvvvv
#<Node 3> has become defeated.
vvvvvvvvvvvvvvvvvvvvvvvvvvvvvvvvvvv
===
Lowering the undefeated-degree-of-support of ((the color of Fred is red) at 50)
retracts the previous answer to #<Query 1: (? x)((the color of Fred is x) at 50)>
===
================ ULTIMATE EPISTEMIC INTERESTS ==================
Interest in (? x)((the color of Fred is x) at 50)
is answered by node 6: ((the color of Fred is blue) at 50)

Now let us return to the problem noted above for PERCEPTUAL-RELIABILITY. This is that we will typically know R-at-t only by inferring it from R-at-t_0 for some $t_0 < t$ (by TEMPORAL-PROJECTION). TEMPORAL-PROJECTION is a backward-reason. That is, given some fact P-at-t, the reasoner only infers P-at-t^* (for $t^* > t$) when that conclusion is of interest. Unfortunately, in PERCEPTUAL-RELIABILITY, R-at-t is not an interest, and so it will not be inferred from R-at-t_0 by TEMPORAL-PROJECTION. This difficulty can be circumvented by formulating PERCEPTUAL-RELIABILITY with an extra forward-premise R-at-t_0 which is marked as a *clue,* and a backward-premise R-at-t:

```
(def-backward-undercutter PERCEPTUAL-RELIABILITY
  :defeatee *perception*
  :forward-premises
    "((the probability of p given ((I have a percept with content p) & R)) ≤ s)"
    (:condition (and (projectible R) (s < 0.99)))
    "(R at time0)"
    (:condition (time0 < time))
    (:clue? t)
  :backward-premises "(R at time)"
  :variables p time R time0 s
  :defeasible? t)
```

The difference between ordinary forward-premises and clues is that when a clue is instantiated by a conclusion that has been drawn, that conclusion is not included in the list of conclusions from which the new conclusion is inferred. The function of clues is only to guide the reasoning. Thus in an application of PERCEPTUAL-RELIABILITY, if R-at-t_0 is concluded, this suggests that R-at-t is true and leads to an interest in it, which can then be inferred from R-at-t_0 by TEMPORAL-PROJECTION.

To illustrate the implementation, consider an example that combines PERCEPTION, TEMPORAL-PROJECTION, and PERCEPTUAL-RELIABILITY (and also a principle about reliable testimony that we have not discussed). At time 1, some object, Fred, looks red to me. I want to know the color of Fred at time 50. By TEMPORAL-PROJECTION, I can infer that Fred is red at time 50. I know that Merrill is a reliable informant, and at time 20 I have a percept of being informed by Merrill that I am wearing blue-tinted glasses. From that I can infer that I am wearing blue-tinted glasses at time 20. At time 30, Fred looks blue to me, from which I infer that Fred has become blue at time 30. From that I infer by TEMPORAL-PROJECTION that Fred is blue at time 50, and hence not red after all. However, wearing blue-tinted glasses reduces the reliability of the judgment that Fred is blue, and by PERCEPTUAL-RELIABILITY, that defeats the inference to that conclusion. This reinstates the conclusion that Fred is red at time 50. Here is a printout of OSCAR performing the same reasoning:

```
================================================================
Inputs:
      (the color of Fred is red) : at cycle 1 with justification 0.8
      (Merrill reports that I am wearing blue tinted glasses) :
            at cycle 20 with justification 1.0
      (the color of Fred is blue) : at cycle 30 with justification 0.8

Given:
      ((the probability of (the color of Fred is blue) given
            ((I have a percept with content
            (the color of Fred is blue)) & I am wearing blue tinted glasses)) ≤ 0.8) :
            with justification = 1.0
      (Merrill is a reliable informant)  :  with justification = 1.0

Ultimate epistemic interests:
      (? x)((the color of Fred is x) at 50)    degree of interest = 0.65
================================================================
THE FOLLOWING IS THE REASONING INVOLVED IN THE SOLUTION
Nodes marked DEFEATED have that status at the end of the reasoning.

   # 1
   ((the probability of (the color of Fred is blue) given ((I have a percept with content (the
   color of Fred is blue)) & I am wearing blue tinted glasses)) ≤ 0.8)
   given
   # 2
   (Merrill is a reliable informant)
   given
                  # 1
                  interest: ((the color of Fred is y) at 50)
                  This is of ultimate interest
||||||||||||||||||||||||||||||||||||||||||||||||||||||||||||||||||||||||||||||||||||||||||||||||||||||||||||||||||
It appears to me that ((the color of Fred is red) at 1)
||||||||||||||||||||||||||||||||||||||||||||||||||||||||||||||||||||||||||||||||||||||||||||||||||||||||||||||||||
   # 3
   It appears to me that ((the color of Fred is red) at 1)
```

4
((the color of Fred is red) at 1)
Inferred by:
    support-link #3 from { 3 } by PERCEPTION
5
((the color of Fred is red) at 50)
Inferred by:
    support-link #4 from { 4 } by TEMPORAL-PROJECTION
    defeaters: { 12 }
This discharges interest 1
    ===
    Justified belief in ((the color of Fred is red) at 50)
    with undefeated-degree-of-support 0.8
    answers #<Query 1: (? x)((the color of Fred is x) at 50)>
    ===
|||||||||||||||||||||||||||||||||||||||||||||||||||||||||||||||||||||||||||||||||||||||||||||||||||||||||||||
It appears to me that ((Merrill reports that I am wearing blue tinted glasses) at 20)
|||||||||||||||||||||||||||||||||||||||||||||||||||||||||||||||||||||||||||||||||||||||||||||||||||||||||||||
6
It appears to me that ((Merrill reports that I am wearing blue tinted glasses) at 20)
7
((Merrill reports that I am wearing blue tinted glasses) at 20)
Inferred by:
    support-link #5 from { 6 } by PERCEPTION
8
(I am wearing blue tinted glasses at 20)
Inferred by:
    support-link #6 from { 2 , 7 } by RELIABLE-INFORMANT
|||||||||||||||||||||||||||||||||||||||||||||||||||||||||||||||||||||||||||||||||||||||||||||||||||||||||||||
It appears to me that ((the color of Fred is blue) at 30)
|||||||||||||||||||||||||||||||||||||||||||||||||||||||||||||||||||||||||||||||||||||||||||||||||||||||||||||
9
It appears to me that ((the color of Fred is blue) at 30)
10
((the color of Fred is blue) at 30)   DEFEATED
Inferred by:
    support-link #7 from { 9 } by PERCEPTION
    defeaters: { 15 } DEFEATED
            # 10
            interest: (((it appears to me that (the color of Fred is blue))
                at 30) @ ((the color of Fred is blue) at 30))
            Of interest as a defeater for support-link 7 for node 10
            This interest is discharged by node 15
            # 11
            interest: (I am wearing blue tinted glasses at 30)
            For interest 10 by PERCEPTUAL-RELIABILITY using
                node 1 with clue 8
            This interest is discharged by node 14
11
((the color of Fred is blue) at 50)   DEFEATED

Inferred by:
        support-link #8 from { 10 } by TEMPORAL-PROJECTION
        defeaters: { 13 }   DEFEATED
This discharges interest 1
===
Justified belief in ((the color of Fred is blue) at 50)
with undefeated-degree-of-support 0.8
answers #<Query 1: (? x)((the color of Fred is x) at 50)>
===
12
~((the color of Fred is red) at 50)   DEFEATED
Inferred by:
        support-link #9 from { 11 } by INCOMPATIBLE-COLORS DEFEATED
defeatees: { link 4 for node 5 }
13
~((the color of Fred is blue) at 50)
Inferred by:
        support-link #10 from { 5 } by inversion from contradictory nodes 12 and 5
defeatees: { link 8 for node 11 }
        vvvvvvvvvvvvvvvvvvvvvvvvvvvvvv
        #<Node 13> has become defeated.
        vvvvvvvvvvvvvvvvvvvvvvvvvvvvvv
        #<Node 5> has become defeated.
        vvvvvvvvvvvvvvvvvvvvvvvvvvvvvv

===
Lowering the undefeated-degree-of-support of ((the color of Fred is red) at 50)
retracts the previous answer to #<Query 1: (? x)((the color of Fred is x) at 50)>
===
14
(I am wearing blue tinted glasses at 30)
Inferred by:
        support-link #11 from { 8 } by TEMPORAL-PROJECTION
This discharges interest 11
15
(((it appears to me that (the color of Fred is blue)) at 30) @ ((the color of Fred is blue) at 30))
Inferred by:
        support-link #12 from { 1 , 14 } by PERCEPTUAL-RELIABILITY
        with clues { 8 }
defeatees: { link 7 for node 10 }
This node is inferred by discharging a link to interest #10
        vvvvvvvvvvvvvvvvvvvvvvvvvvvvvv
        #<Node 10> has become defeated.
        vvvvvvvvvvvvvvvvvvvvvvvvvvvvvv
        #<Node 11> has become defeated.
        vvvvvvvvvvvvvvvvvvvvvvvvvvvvvv
        #<Node 12> has become defeated.
        vvvvvvvvvvvvvvvvvvvvvvvvvvvvvv

===
Justified belief in ((the color of Fred is red) at 50)

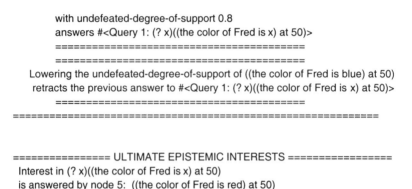

with undefeated-degree-of-support 0.8
answers #<Query 1: (? x)((the color of Fred is x) at 50)>
==
==
Lowering the undefeated-degree-of-support of ((the color of Fred is blue) at 50)
retracts the previous answer to #<Query 1: (? x)((the color of Fred is x) at 50)>
==
===

=============== ULTIMATE EPISTEMIC INTERESTS ==================
Interest in (? x)((the color of Fred is x) at 50)
is answered by node 5: ((the color of Fred is red) at 50)

It may be illuminating to draw the inference-graph for this problem, as in figure 7.6.

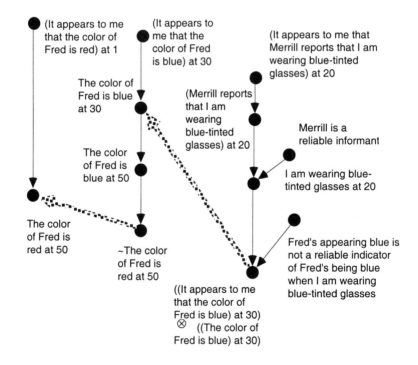

Figure 7.6 Inference-graph

7. Reasoning about Change

In the real world, things change. A cognizer residing in a changing world must be able to reason about change and persistence. This requires four kinds of reasoning. First, the cognizer must be able to dip into the world perceptually, acquiring information about the current state of the world. Second, she must be able to combine information obtained from different perceptual excursions and, through inference, construct a coherent picture of the world. Third, she must be able to detect changes perceptually and update her picture of the world accordingly. Fourth, she must be able to acquire general causal information about "how the world works" and use that to predict the results of changes either observed by her or wrought by her own actions. We have addressed the first three kinds of reasoning above. Now let us turn to the fourth.

The problem of explaining how causal reasoning works turns out to be more difficult than philosophers originally supposed. The difficulties were first noted by AI researchers working on planning theory. Planning theory is concerned with the construction of automated systems that will produce plans for the achievement of specified goals. In order to construct a plan, a cognizer must be able to predict the outcomes of the various actions that a plan might prescribe. For this purpose, let us suppose the cognizer has all the general background knowledge she might need. Consider a very simple planning problem. The cognizer is standing in the middle of a room, and the light is off. The light switch is by the door. The cognizer wants the light to be on. The obvious plan for achieving this goal is to walk to the vicinity of the light switch and activate the switch. *We human beings* can see immediately that, barring unforeseen difficulties, this is a good plan for achieving the goal. If an artificial agent is to be able to see this as well, it must be able to infer that the execution of this plan will, barring unforeseen difficulties, achieve the goal. The reasoning required seems easy. First, the switch is observed to be at position S. Our background knowledge allows us to infer that if we walk towards position S, we will shortly be in that vicinity. Second, our background knowledge allows us to infer that when we are in the vicinity of the switch, we can activate it. Third, it informs us that when we activate the switch, the light will come on. It may seem that this information is all that is required to conclude that if the plan is executed then the light will come on. But in fact, one more premise is required. We know that the switch is initially at position S. However, for the plan to work, we must know that the switch will still be at position S when we get there. In other words, we have to know that walking to position S does not change the position of the switch. This of course is something that we do know, but what this example illustrates is that reasoning about what will change if an action is performed or some other event occurs generally presupposes knowing what will not change.

Early attempts in AI to model reasoning about change tried to do so deductively by formulating axioms describing the environment in which the planner was operating and then using those axioms to deduce the outcomes of executing proposed plans. The preceding example illustrates that among the axioms describing the environment there must be both causal axioms about the effects of various actions or events under specified circumstances, and a number of axioms about what does not change when actions are performed or events occur under specified circumstances. The latter axioms were called *Frame Axioms*.[176] In our simple example, we can just add a frame axiom to the effect that the switch will still be at position S if the agent walks to that position, and then the requisite reasoning can be performed. However, in pursuing this approach, it soon became apparent that more complicated situations required vastly more (and more complicated) frame axioms. A favorite example of early AI researchers was the Blocks World, in which children's building blocks are scattered about and piled on top of each other in various configurations, and the planning problem is to achieve a certain configuration of blocks. If we make the world sufficiently simple, then we can indeed axiomatize it and reason about it deductively. But if we imagine a world whose possibilities include all the things that can happen to blocks in the real world, this approach becomes totally impractical. For instance, moving a block does not normally change its color. But it might if, for example, an open can of paint is balanced precariously atop the block. If we try to apply the axiomatic approach to real-world situations, we encounter three problems. First, in most cases we will be unable to even formulate a suitable set of axioms that does justice to the true complexity of the situation. But second, even if we could, we would find it necessary to construct an immense number of extraordinarily complex frame axioms. And third, if we then fed these axioms to an automated reasoner and set it the task of deducing the outcomes of a plan, the reasoner would be forced to expend most of its resources reasoning about what does not change rather than what does change, and it would quickly bog down and be unable to draw the desired conclusions about the effects of executing the plan.[177]

The upshot of this is that in realistically complicated situations, ax-iomatizing the situation and reasoning about it deductively is made un-manageable by the proliferation and complexity of frame axioms. What became known as the *Frame Problem* is the problem of reorganizing rea-soning about change so that reasoning about non-changes can be done efficiently.[178]

It has often gone unappreciated that the Frame Problem is equally a problem for human epistemology. Humans can perform the requisite

---

176. John McCarthy and Patrick Hayes (1969).

177. .On this last point, the reader who lacks experience with automated reasoners will have to take our word, but this is a point about which no AI practitioner would disagree.

178. John McCarthy and Patrick Hayes (1969), Lars-Erik Janlert (1987).

reasoning, so they instantiate a solution to the Frame Problem. However, it is not obvious how they do it, any more than it is obvious how they perform inductive reasoning or probabilistic reasoning or any other epistemologically problematic species of reasoning. Describing such reasoning is a task for epistemology. Furthermore, it seems quite likely that the best way to solve the Frame Problem for artificial rational agents is to figure out how it is solved in human reasoning and then implement that solution in artificial agents. Thus the epistemological problem and the AI engineering problem become essentially the same problem.

The Frame Problem arose in the context of an attempt to reason about persistence and change deductively. That may seem naive in contemporary epistemology, but it should be borne in mind that at the time this work was taking place (the late 60's), philosophy itself was just beginning to appreciate the necessity for nondeductive reasoning, and at that time the predominant view was still that good arguments must be deductively valid. Thirty years later, nobody believes that. Some kind of defeasible reasoning is recognized as the norm, with deductive reasoning being the exception. To what extent does the Frame Problem depend upon its deductivist origins?

This same question occurred to AI researchers. Several authors proposed eliminating frame axioms altogether by reasoning about change defeasibly and adopting some sort of defeasible inference scheme to the effect that it is reasonable to believe that something doesn't change unless you are forced to conclude otherwise.[179] The temporal projection principles formulated in section six can be regarded as a precise formulation of the defeasible inference schemes sought. Unfortunately, these principles do not solve the Frame Problem.

Steve Hanks and Drew McDermott (1986) were the first to observe that even with defeasible principles of persistence, a reasoner will often be unable to determine what changes and what does not. They illustrated this with what has become known as "the Yale shooting problem". The general form of the problem is this. Suppose we have a causal law to the effect that if P is true at a time t and action A is performed at that time, then Q will be true shortly thereafter. (More generally, A could be anything that becomes true at a certain time. What is significant about actions is that they are changes.) Suppose we know that P is true now, and Q false. What should we conclude about the results of performing action A in the immediate future? Hanks and McDermott illustrate this by taking P to be "The gun is loaded and pointed at Jones", Q to be "Jones is dead", and A to be the action of pulling the trigger. We suppose (simplistically) that there is a causal law dictating that if the trigger is

---

179. Erik Sandewall (1972), Drew McDermott (1982), John McCarthy (1986). This played an important role in motivating research in AI on defeasible reasoning and nonmonotonic logic. For example, see the collection of papers in Mathew Ginsberg (1987).

pulled on a loaded gun that is pointed at someone, that person will shortly be dead. Under these circumstances, it seems clear that we should conclude that Jones will be dead shortly after the trigger is pulled.

Without a principle of temporal projection, all we can infer from what we are given is that when A is performed either P will no longer be true or Q will be true shortly thereafter. Intuitively, we want to conclude (at least defeasibly) that P will remain true at the time A is performed and Q will therefore become true shortly thereafter. It might seem that TEMPORAL-PROJECTION provides the needed extra step. Because P is now true, we have a defeasible reason for believing that P will still be true when A is performed, and from that we can infer that Q will become true shortly thereafter. Unfortunately, TEMPORAL-PROJECTION licenses another inference as well. It provides a defeasible reason for believing that, because Q is initially false, it will still be false after A is performed. This is diagrammed in figure 7.7. We know that one of these defeasible conclusions will be false, but we have no basis for choosing between them, so this becomes a case of collective defeat. That, however, is the intuitively wrong answer.

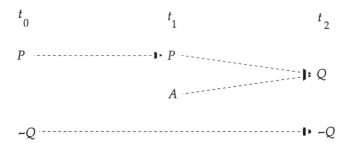

Figure 7.7 The Yale shooting problem

When we reason about causal mechanisms, we think of the world as "unfolding" temporally, and changes only occur when they are forced to occur by what has already happened. In our example, when A is performed, nothing has yet happened to force a change in P, so we conclude defeasibly that P remains true. But given the truth of P, we can then deduce that at a slightly later time, Q will become true. Thus when causal mechanisms force there to be a change, we conclude defeasibly that the change occurs in the later states rather than the earlier states. This seems to be part of what we mean by describing something as a causal mechanism. Causal mechanisms are systems that force changes, where "force" is to be understood in terms of temporal unfolding.[180]

---

180. This intuition is reminiscent of Yoav Shoham's (1987) "logic of chronological

When reasoning about such a causal system, part of the force of describing it as causal must be that the defeasible presumption against the effect occurring is somehow removed. Thus, although we normally expect Jones to remain alive, we do not expect this any longer when he is shot. To remove a defeasible presumption is to defeat it. This suggests that there is some kind of general "causal" defeater for the temporal projection principles adumbrated above. The problem is to state this defeater precisely. As a first approximation we might try:

(7.3)   For every $\varepsilon \geq 0$ and $\delta > 0$, "$A\&P$-at-$(t+\varepsilon)$ & $(A\&P$ causes $Q)$" is an undercutting defeater for the defeasible inference from $\sim Q$-at-t to $\sim Q$-at-$(t+\varepsilon+\delta)$ by TEMPORAL-PROJECTION.

The temporal-unfolding view of causal reasoning requires causation to be temporally asymmetric. That is, "$A\&P$ causes Q" means, in part, that if $A\&P$ becomes true then Q will *shortly* become true. This precludes *simultaneous* causation, in which Q is caused to be true at t by $A\&P$ being true at t, because in such a case temporal ordering would provide no basis for preferring the temporal projection of P over that of $\sim Q$. This may seem problematic, on the grounds that simultaneous causation occurs throughout the real world. For instance, colliding billiard balls in classical physics might seem to illustrate simultaneous causation. However, this is a mistake. If two billiard balls collide at time t with velocity vectors pointing towards each other, they do not also have velocity vectors pointing away from each other at the very same time. Instead, this illustrates what Pollock (1984) called *instantaneous* causation. Instantaneous causation requires that if $A\&P$ becomes true at t, then Q becomes true *immediately afterwards*, i.e., for some $\delta > 0$, Q will be true throughout the clopen interval $(t, t+\delta]$.[181] Instantaneous causation is all that is required for describing the real world.

Principle (7.3) was formulated in terms of causation. However, that introduces unnecessary complexities. For example, it is generally assumed in the philosophical literature on causation that if P causes Q then Q would not have been true if P were not true.[182] This has the consequence that when there are two independent factors each of which would be sufficient by itself to cause the same effect, if both occur then neither causes it. These are cases of causal overdetermination. A familiar example of causal overdetermination occurs when two assailants shoot a common

---

ignorance", although unlike Shoham, we propose to capture the intuition without modifying the structure of the system of defeasible reasoning. This is also related to the proposal of Michael Gelfond and Vladimir Lifschitz (1993), and to the notion of progressing a database discussed in Fangzhen Lin and Raymond Reiter (1994 and 1995).

181. A clopen interval $(x, y]$ consists of all real numbers z such that $x < z \leq y$. We assume that time has the structure of the reals, although that assumption is not required for the implementation.

182. See David Lewis (1973). See Pollock (1984) for more details about the relationship between causes and counterfactual conditionals.

victim at the same time. Either shot would be fatal. The result is that neither shot is such that if it had not occurred then the victim would not have died, and hence, it is generally maintained, neither shot caused the death of the victim. However, this kind of failure of causation ought to be irrelevant to the kind of causal reasoning under discussion in connection with change. Principle (7.3) ought to apply to cases of causal overdetermination as well as to genuine cases of causation. This indicates that the intricacies of the analysis of "cause" are irrelevant in the present context.

We follow Pollock (1984) in assuming that all varieties of causation (including causal overdetermination) arise from the instantiation of "causal laws". Causal laws are expressed by *nomic generalizations*, which are discussed at length in Pollock (1989). Nomic generalizations are symbolized as "$P \Rightarrow Q$", where P and Q are formulas and '$\Rightarrow$' is a variable-binding operator, binding all free occurrences of variables in P and Q. An informal gloss on "$P \Rightarrow Q$" is "Any physically-possible P would be a Q". For example, the law that electrons are negatively charged could be written "(x is an electron) $\Rightarrow$ (x is negatively charged)".

A rule of universal instantiation applies to nomic generalizations, allowing us to derive less general nomic generalizations:

If 'x' is free in P and Q, and $P(x/a)$ and $Q(x/a)$ result from substituting the constant term 'a' for 'x', then $(P \Rightarrow Q)$ entails $(P(x/a) \Rightarrow Q(x/a))$.

We propose to replace "($A\&P$ causes Q)" in (7.3) by "($A\&P \Rightarrow Q$ will shortly be true)", where the latter typically results from instantiating more general laws. More precisely, let us define "A when P is causally sufficient for Q after an interval ε" to mean

$$(\forall t)\{(A\text{-}at\text{-}t \;\&\; P\text{-}at\text{-}t) \Rightarrow (\exists \delta)Q\text{-}throughout\text{-}(t+\varepsilon \,,\, t+\varepsilon+\delta]\}.$$

Instantaneous causation is causal sufficiency with an interval 0.

Our proposal is to replace "causes" by "causal sufficiency" in (7.3). Modifying it to take account of the interval over which the causation occurs:

CAUSAL-UNDERCUTTER
Where $t_0 \leq t_1$ and $(t_1+\varepsilon) < t$, "$A\text{-}at\text{-}t_1$ & $Q\text{-}at\text{-}t_1$ & (A when Q is causally sufficient for $\sim P$ after an interval ε)" is a defeasible undercutting defeater for the inference from $P\text{-}at\text{-}t_0$ to $P\text{-}at\text{-}t$ by TEMPORAL-PROJECTION.

CAUSAL-UNDERCUTTER uses causal connections to defeat applications of TEMPORAL-PROJECTION. For causal reasoning, we also want to use causal connections to support inferences about what will happen. This is more complicated than it might initially seem. The difficulty is that, for example,

the gun is fired when the gun is loaded is causally sufficient for ~(Jones is alive) after an interval 20 does not imply that if the gun is fired at t and the gun is loaded at t then Jones is dead at $t+20$. Recall the discussion of instantaneous causation. All that is implied is that Jones is dead over some interval open on the left and with $t+20$ as the lower bound. We can conclude that *there is* a time after $t+20$ at which Jones is dead, but it does not follow as a matter of logic that Jones is dead at any particular time because, at least as far as this causal law is concerned, Jones could become alive again after becoming dead. To infer that Jones is dead at a particular time after $t+20$, we must combine the causal sufficiency with temporal projection. This yields the following principle:

CAUSAL-IMPLICATION
If Q is temporally projectible and $(t+\varepsilon) < t^*$, then "$(A$ when P is causally sufficient for Q after an interval $\varepsilon)$ & A-at-t & P-at-t" is a defeasible reason for "Q-at-t^*".

These principles can be illustrated by applying them to the Yale Shooting Problem. That problem arises from the fact that if we have CAUSAL-IMPLICATION but do not have CAUSAL-UNDERCUTTER, then it cannot be inferred that Jones is dead after the shooting. Instead we get collective defeat, as indicated in figure 7.8.

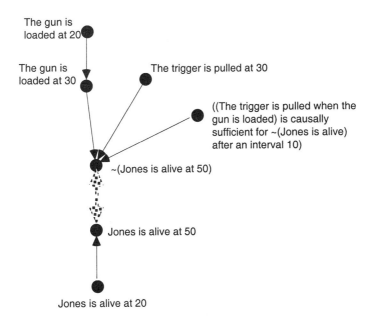

Figure 7.8 The unsolved Yale shooting problem

On the other hand, if we allow the reasoner to use CAUSAL-UNDERCUTTER, it is able to conclude that Jones is dead after the shooting:

```
================================================================
Given:
      ((Jones is alive) at 20)  :  with justification = 1.0
      (the gun is loaded at 20)  :  with justification = 1.0
      (the gun is fired at 30)  :  with justification = 1.0
      (the gun is fired when the gun is loaded is causally sufficient for
        ~(Jones is alive) after an interval 10)  :  with  justification = 1.0

Ultimate epistemic interests:
   (? (Jones is alive) at 50)    degree of interest = 0.75
================================================================
```

THE FOLLOWING IS THE REASONING INVOLVED IN THE SOLUTION
Nodes marked DEFEATED have that status at the end of the reasoning.

\# 1
(the gun is loaded at 20)
given
\# 2
((Jones is alive) at 20)
given
\# 3
(the gun is fired at 30)
given
\# 4
(the gun is fired when the gun is loaded is causally sufficient for ~(Jones is alive) after an interval 10)
given

> \# 1
> interest: ((Jones is alive) at 50)
> This is of ultimate interest
> For interest 9 by neg-at-intro
> \# 2
> interest: ~((Jones is alive) at 50)
> This is of ultimate interest
> Of interest as a defeater for support-link 5 for node 5
> \# 3
> interest: (~(Jones is alive) at 50)
> For interest 2 by NEG-AT-INTRO
> This interest is discharged by node 7

\# 5
((Jones is alive) at 50)   DEFEATED
Inferred by:
      support-link #5 from { 2 } by TEMPORAL-PROJECTION
      defeaters: { 10 , 8 } DEFEATED

This discharges interest 1
> # 4
> interest: (((Jones is alive) at 20) @ ((Jones is alive) at 50))
> Of interest as a defeater for support-link 5 for node 5

==
Justified belief in ((Jones is alive) at 50)
with undefeated-degree-of-support 0.941
answers #<Query 1: (? ((Jones is alive) at 50))>
==

> # 5
> interest: (the gun is loaded at 30)
> For interest 3 by CAUSAL-IMPLICATION
> For interest 4 by CAUSAL-UNDERCUTTER
> This interest is discharged by node 6

6
(the gun is loaded at 30)
Inferred by:
> support-link #6 from { 1 } by TEMPORAL-PROJECTION

This discharges interest 5
7
(~(Jones is alive) at 50)
Inferred by:
> support-link #7 from { 4 , 3 , 6 } by CAUSAL-IMPLICATION
> defeaters: { 9 }

This node is inferred by discharging interest #3
> # 9
> interest: ~(~(Jones is alive) at 50)
> Of interest as a defeater for support-link 7 for node 7

8
~((Jones is alive) at 50)
Inferred by:
> support-link #8 from { 7 } by NEG-AT-INTRO

defeatees: { link 5 for node 5 }
This node is inferred by discharging interest #2
9
~(~(Jones is alive) at 50)   DEFEATED
Inferred by:
> support-link #9 from { 5 } by inversion from contradictory nodes 8 and 5
> DEFEATED

defeatees: { link 7 for node 7 }
10
(((Jones is alive) at 20) ⊗ ((Jones is alive) at 50))
Inferred by:
> support-link #10 from { 4 , 3 , 6 } by CAUSAL-UNDERCUTTER

defeatees: { link 5 for node 5 }
This node is inferred by discharging interest #4

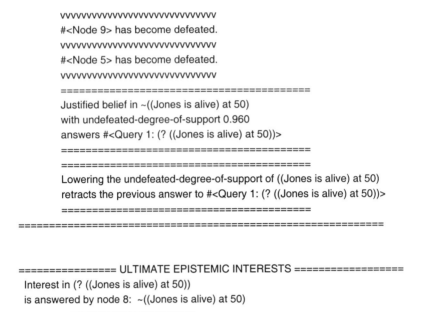

vvvvvvvvvvvvvvvvvvvvvvvvvvvvvvvvvvv
#<Node 9> has become defeated.
vvvvvvvvvvvvvvvvvvvvvvvvvvvvvvvvvvv
#<Node 5> has become defeated.
vvvvvvvvvvvvvvvvvvvvvvvvvvvvvvvvvvv
===
Justified belief in ~((Jones is alive) at 50)
with undefeated-degree-of-support 0.960
answers #<Query 1: (? ((Jones is alive) at 50))>
=======================================
=======================================
Lowering the undefeated-degree-of-support of ((Jones is alive) at 50)
retracts the previous answer to #<Query 1: (? ((Jones is alive) at 50))>
=======================================
===

================ ULTIMATE EPISTEMIC INTERESTS ==================
Interest in (? ((Jones is alive) at 50))
is answered by node 8: ~((Jones is alive) at 50)

This reasoning produces the inference-graph of figure 7.9.

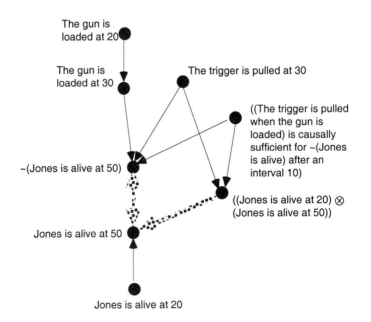

Figure 7.9 The solved Yale shooting problem

Thus the Yale Shooting Problem is solved. We propose that the same epistemic machinery that solves this problem constitutes a general solution to the Frame Problem. In light of our previous remarks about toy problems, one should be wary of generalizing too quickly from a single example. Comfortably ensconced in our armchairs, we may agree that we see no reason why the machinery that solves this problem should not work in general. But to be confident of such a claim, one must test it on a variety of problems of realistic complexity. In this respect, AI and epistemology are experimental sciences. We have tested the proposal on some more complicated problems,[183] where it performs successfully, and we invite the reader to download the code and test it on further problems.

8. The Statistical Syllogism

Our ability to maneuver our way successfully through a complex environment depends heavily on our being able to predict what will happen under various circumstances. This requires us to have general knowledge of the world. When we have knowledge of exceptionless generalizations of the form "All F's are G's", then all it takes is simple deduction to infer that a particular F will be a G. However, we do not believe very many exceptionless generalizations. Most of the generalizations we believe and in terms of which we guide our behavior are statistical, of the form "Most F's are G's". In order to function in a world like ours, a rational agent must be equipped with rules (1) enabling it to form beliefs in statistical generalizations, and (2) enabling it to make inferences from those statistical generalizations to beliefs about individual matters of fact. Most inferences of the latter sort proceed in terms of the *statistical syllogism*. The non-numerical version of the statistical syllogism can be formulated roughly as follows:

Most F's are G's.
This is an F.
This is a G.

We often reason in roughly this way. For instance, on what basis do I believe what I read in the newspaper? Certainly not that everything printed in the newspaper is true. No one believes that. But I do believe that most of what is printed in the newspaper is true, and that justifies me in believing individual newspaper reports. What "Most F's are G's" requires is that the probability is high of an arbitrary G's being an F. Accordingly, the statistical syllogism can be rewritten as:

---

183. For an extended example, see Pollock (1998).

prob(G/F) $\geq r$.
This is an F.
This is a G.

The choice of r will vary from one situation to another, depending upon the degree of certainty that is demanded.

Obviously, the inference described by the statistical syllogism is a defeasible one. This suggests that the statistical syllogism could be formulated more precisely as follows:

(7.4)   If $r > 0.5$ then "Fc and prob(G/F) $\geq r$" is a defeasible reason for "Gc", the strength of the reason depending upon the value of r.

It is illuminating to consider how this inference-scheme handles a well known paradox. This is the *lottery paradox*, due to Henry Kyburg (1961). Suppose you hold one ticket in a fair lottery consisting of 1 million tickets, and suppose it is known that one and only one ticket will win. Observing that the probability is only .000001 of a ticket's being drawn given that it is a ticket in the lottery, it seems reasonable to accept the conclusion that your ticket will not win. But by the same reasoning, it will be reasonable to believe, for each ticket, that it will not win. However, these conclusions conflict jointly with something else we are justified in believing, namely, that some ticket will win. So what should we believe?

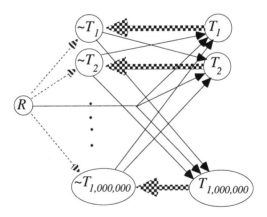

Figure 7.10 The lottery paradox

Principle (7.4) provides an answer to this question. The reasoning is described in figure 7.10. We begin with a description R of the lottery. From that we construct a defeasible inference to each conclusion of the form "Ticket n will not be drawn", symbolized '$\sim T_n$'. However, R entails that some ticket will be drawn, i.e., it entails the disjunction $T_1 \vee T_2 \vee \ldots \vee$

$T_{1,000,000}$. For each n, if we combine the arguments for all $\sim T_i$ for $i \neq n$ with the argument for the disjunction $T_1 \vee T_2 \vee \ldots \vee T_{1,000,000}$, we get an argument for T_n. This constitutes a rebutting defeater for the argument for $\sim T_n$. The result is a case of collective defeat. For each n, there is a status-assignment assigning "undefeated" to all $\sim T_i$ for $i \neq n$, and hence assigning "undefeated" to T_n and "defeated" to $\sim T_n$. So all of the conclusions to the effect that a particular ticket will not be drawn are defeated. This is a case of collective defeat. We can still, of course, conclude of each ticket that it will *probably* not be drawn, but we cannot detach the probability and conclude that it will definitely not be drawn. After all, one of them will be drawn, and we do not know which.

Although principle (7.4) resolves the lottery paradox by making it a case of collective defeat, it turns out that the very fact that (7.4) can handle the lottery paradox in this way shows that it cannot be correct. The difficulty is that every case of high probability can be recast in a form that makes it similar to the lottery paradox. We need only assume that $\mathrm{prob}(G/F) < 1$. Pick the smallest integer n such that $\mathrm{prob}(G/F) < 1 - 1/2^n$. Now consider n fair coins, unrelated to each other and unrelated to c's being F or G. Let T_i be "is a toss of coin i" and let H be "is a toss that lands heads". There are 2^n *Boolean conjunctions* of the form "$(\sim)Hx_1$ & ... & $(\sim)Hx_n$" where each tilde in parentheses can be either present or absent. For each Boolean conjunction $\beta_j x_1 \ldots x_n$,

$$\mathrm{prob}(\beta_j x_1 \ldots x_n / T_1 x_1 \& \ldots \& T_n x_n) = 2^{-n}.$$

Consequently, because the coins were chosen to be unrelated to F and G,

$$\mathrm{prob}(\sim\beta_j x_1 \ldots x_n / Fx \& T_1 x_1 \& \ldots \& T_n x_n) = 1 - 2^{-n}.$$

By the probability calculus, a disjunction is at least as probable as its disjuncts, so

$$\mathrm{prob}(\sim Gx \vee \sim\beta_j x_1 \ldots x_n / Fx \& T_1 x_1 \& \ldots \& T_n x_n) \geq 1 - 2^{-n} > \mathrm{prob}(Gx/Fx).$$

Let $t_1, \ldots, t_n$ be a sequence consisting of one toss of each coin. As we know "$Fc \& T_1 t_1 \& \ldots \& T_n t_n$", (7.4) gives us a defeasible reason for believing each disjunction of the form

$$\sim Gc \vee \sim\beta_j t_1 \ldots t_n.$$

By the propositional calculus, the set of all these disjunctions is equivalent to, and hence entails, $\sim Gc$. Thus we can construct an argument for $\sim Gc$ in which the only defeasible steps involve the use of (7.4) in connection with probabilities at least as great as that used in defending Gc. Hence, we have a situation formally identical to the lottery paradox. Therefore, principle (7.4) makes this a case of collective defeat as well, with the consequence that if prob(G/F) has any probability less than 1, we cannot use (7.4) to draw any justified conclusion from this high probability.

The difficulty can be traced to the assumption that F and G in (7.4) can be arbitrary formulas. Basically, we need a constraint that, when applied to the above argument, precludes applying (7.4) to the disjunctions "$\sim Gc \vee \sim\beta_j t_1 ... t_n$". It turns out that disjunctions create repeated difficulties throughout the theory of probabilistic reasoning. This is easily illustrated in the case of (7.4). For instance, it is a theorem of the probability calculus that prob($F/G \vee H$) ≥ prob(F/G)·prob($G/G \vee H$). Consequently, if prob(F/G) and prob($G/G \vee H$) are sufficiently large, it follows that prob($F/G \vee H$) ≥ r. For example, because the vast majority of birds can fly and because there are many more birds than giant sea tortoises, it follows that most things that are either birds or giant sea tortoises can fly. If Herman is a giant sea tortoise, (7.4) would give us a reason for thinking that Herman can fly, but notice that this is based simply on the fact that most birds can fly, which should be irrelevant to whether Herman can fly. This example indicates that arbitrary disjunctions cannot be substituted for G in (7.4).

Nor can arbitrary disjunctions be substituted for F in (7.4). By the probability calculus, prob($F \vee G/H$) ≥ prob(F/H). Therefore, if prob(F/H) is high, so is prob($F \vee G/H$). Thus, because most birds can fly, it is also true that most birds can either fly or swim the English Channel. By (7.4), this should be a reason for thinking that a starling with a broken wing can swim the English Channel, but obviously it is not.

There must be restrictions on the properties F and G in (7.4). The properties that cause trouble turn out to be the same as the ones that cause trouble in induction, so the constraint we need is the same, viz., projectibility. We can, accordingly, formulate a correct principle of the statistical syllogism by adding a projectibility constraint to (7.4):

STATISTICAL-SYLLOGISM
If G is projectible with respect to F and $r > 0.5$, then "Fc & prob(G/F) ≥ r" is a defeasible reason for believing "Gc", the strength of the reason depending upon the value of r.

As noted earlier, the term "projectible" comes from the literature on induction. Nelson Goodman (1955) was the first to observe that principles of induction require a projectibility constraint. There are two reasons for thinking that the same constraint is required for both principles of induction and the statistical syllogism. First, there is the purely empirical observation that the same predicates seem to cause trouble in both cases. The worst culprits are disjunctions. There is also a theoretical reason for identifying the two constraints. It was shown in Pollock (1989) that familiar-looking principles of induction can actually be derived from STATISTICAL-SYLLOGISM and the probability calculus, with the result that the projectibility constraint on induction turns out to derive from the projectibility constraint on STATISTICAL-SYLLOGISM. It is the same property of projectibility that is involved in both cases.

The reason provided by STATISTICAL-SYLLOGISM is only a defeasible reason. Like any other defeasible reason, it can be defeated by having a reason for denying the conclusion. The reason for denying the conclusion constitutes a rebutting defeater. But there are also important undercutting defeaters for STATISTICAL-SYLLOGISM. In STATISTICAL-SYLLOGISM, we infer the truth of "Gc" on the basis of probabilities conditional on a limited set of facts about c (the facts expressed by "Fc"). But if we know additional facts about c that alter the probability, that defeats the defeasible reason:

SUBPROPERTY-DEFEAT
If G is projectible with respect to H then "Hc & prob($G/F\&H$) $\neq$ prob(G/F)" is an undercutting defeater for STATISTICAL-SYLLOGISM.

These are called *subproperty defeaters*. SUBPROPERTY-DEFEAT amounts to a kind of "total evidence requirement". It requires us to make our inference on the basis of the most comprehensive facts regarding which we know the requisite probabilities.[184] To illustrate, suppose we know that Boris' fingers are covered with spots, and it is extremely probable that if a person's fingers are covered with spots then they are suffering from myometatarsilitis. Then we have a reason for believing that Boris is suffering from myometatarsilitis. But suppose we also know that Boris does not have a fever, and the probability is low that a person has myometatarsilitis if his fingers are covered with spots but he does not have a fever. Then a second application of statistical syllogism gives us a reason for thinking that Boris is not suffering from myometatarsilitis. Thus we have two defeasible arguments with conflicting conclusions, as diagrammed in figure 7.11. If the reasoning stopped here, this would be a case of collective defeat—each inference provides a rebutting defeater for the other, and no way of deciding between them has been provided. But intuitively, we want to give precedence to the second inference because it is based upon more information. This is accomplished by having a subproperty defeater. We can then draw the undefeated conclusion that Boris does not have myometatarsilitis.

---

184. The need for subproperty defeaters was first observed in Pollock (1983a). David Touretzky (1984) subsequently introduced similar defeaters for use in defeasible inheritance hierarchies.

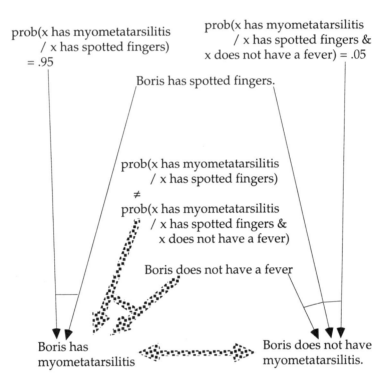

Figure 7.11 Subproperty defeat

9. Induction

Much of our reasoning turns upon having generalizations at our disposal. The generalizations can be either exceptionless generalizations, of the form "All A's are B's", or statistical generalizations to the effect that the probability of an A being a B is high. Note that the latter is an indefinite probability. We typically become justified in believing generalizations of either sort by reasoning inductively, although the kinds of induction that support the two kinds of generalization are somewhat different.

The simplest kind of induction is *enumerative induction*, which proceeds from the observation that all members of a sample X of A's are B's, and makes a defeasible inference to the conclusion that all A's are B's. This is just the Nicod Principle, suitably restricted with a projectibility constraint:

(7.5)   If B is projectible with respect to A, then "X is a sample of A's all of which are B's" is a defeasible reason for "All A's are B's".

However, formulating the principle in this way merely scratches the surface of the logical complexities associated with enumerative induction. First, it is difficult to give a satisfactory account of the defeaters for this defeasible reason. Of course, counterexamples to the generalization are automatically rebutting defeaters, but the more interesting question is to ask what the undercutting defeaters look like. An important variety of undercutting defeater charges that X is not a "fair sample". We illustrated this earlier with the example of the pollster attempting to predict what proportion of residents of Indianapolis will vote for the Republican gubernatorial candidate in the next election. Fair sample defeaters are easy to illustrate but hard to characterize in a general fashion. It is clear that they must be undercutting defeaters rather than a rebutting defeaters, but how exactly to formulate them is not at all clear.[185]

Another source of complexity concerns the strength of an inductive reason. As a general rule, the larger the sample, the stronger the reason. A separate factor concerns the diversity of the sample. For instance, if we are attempting to confirm a generalization about all mammals, our confirmation will be stronger if we sample wild animals in addition to domesticated animals.[186] It is not at all clear how to combine all of this into a precise principle telling us how strong the inductive reason is.

Enumerative induction has been a favorite topic of philosophers, but *statistical induction* is much more important for the construction of a rational agent. It is rare that we are in a position to confirm exceptionless universal generalizations. Induction usually leads us to statistical generalizations that either estimate the probability of an A being a B, or the proportion of actual A's that are B's, or more simply, may lead us to the conclusion that most A's are B's. Such statistical generalizations are very useful, because the statistical syllogism enables a cognizer to make defeasible inferences from them to non-probabilistic conclusions.

A very rough formulation of a principle of statistical induction would proceed as follows:

(7.6)   If B is projectible with respect to A, then "X is a sample of n A's r of which are B's" is a defeasible reason for "prob(B/A) is approximately equal to r/n".

---

185. Fair sample defeaters are discussed at greater length in Pollock (1989), but no general account is given.

186. Further discussion of this can be found in Pollock (1989, 315ff).

All of the problems that arise for a precise theory of enumerative induction arise again for statistical induction. In addition, there is a further problem connected with the fact that the conclusion of (7.6) is only that the probability is *approximately* equal to r/n. How good should we expect the approximation to be? There are obvious things to say, such as that the degree of approximation should improve as the size of the sample increases. It was argued in Pollock (1989) that this can be made precise as follows. Abbreviating r/n as f, define:

(7.7) $L(n,r,p) = (p/f)^{nf} \cdot ((1-p)/(1-f))^{n(1-f)}$.

$L(n,r,\text{p})$ is the *likelihood ratio* of "prob$(A/B) = p$" to "prob$(A/B) = f$". It was argued that each degree of justification corresponds to a minimal likelihood ratio, so we can take the likelihood ratio to be a measure of the degree of justification. For each likelihood ratio α we obtain the α-*rejection class* R_α and the α-*acceptance class* A_α:

(7.8) $R_\alpha = \{p \mid L(n,r,p) \le \alpha\}$

(7.9) $A_\alpha = \{p \mid L(n,r,p) > \alpha\}$.

We are justified to degree α in concluding that prob(A/B) is not a member of R_α, and hence we are justified to degree α in believing that prob(A/B) is a member of A_α. If we plot the likelihood ratios, we get a bell curve centered around r/n, with the result that A_α is an interval around r/n and R_α consists of the tails of the bell curve (figure 7.12). Low likelihood ratios correspond to a high degree of justification for *rejecting* that value for prob(A/B), and so the region around r/n consists of those values we cannot reject, that is, it consists of those values that might be the actual value. This provides us with justification for believing that prob(A/B) lies in a precisely defined interval around the observed relative frequency, the width of the interval being a function of the degree of justification. For illustration, some typical values of the acceptance interval are listed in table 7.1. Reference to the acceptance level reflects the fact that attributions of justification are relative to an index measuring the requisite degree of justification. Sometimes an acceptance level of .1 may be reasonable, at other times an acceptance level of .01 may be required, and so forth.

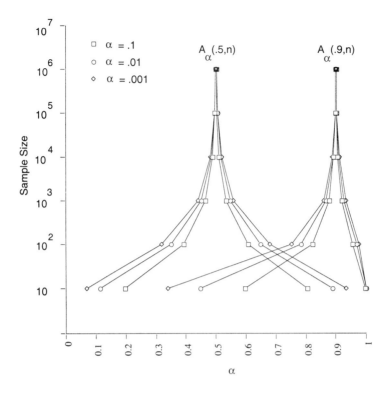

Figure 7.12 Acceptance intervals

The "justification" of principles of induction is one of the traditional problems of philosophy, but in an important sense, induction needs no justification. It is an irreducible constituent of our procedural knowledge for how to cognize, and is accordingly a primitive part of any correct theory of human rationality. This is not to say, however, that principles of induction are philosophically trouble free. Major questions remain about the precise form of the epistemic norms governing induction. Suggestions have been made here about some of the details of those epistemic norms, but we do not yet have a complete account.

Table 1. Values of $A_\alpha(f,n)$.

$A_\alpha(.5,n)$
n

α	10	10^2	10^3	10^4	10^5	10^6
.1	[.196,.804]	[.393,.607]	[.466,.534]	[.489,.511]	[.496,.504]	[.498,.502]
.01	[.112,.888]	[.351,.649]	[.452,.548]	[.484,.516]	[.495,.505]	[.498,.502]
.001	[.068,.932]	[.320,.680]	[.441,.559]	[.481,.519]	[.494,.506]	[.498,.502]

$A_\alpha(.9,n)$
n

α	10	10^2	10^3	10^4	10^5	10^6
.1	[.596,.996]	[.823,.953]	[.878,.919]	[.893,.907]	[.897,.903]	[.899,.901]
.01	[.446,1.00]	[.785,.967]	[.868,.927]	[.890,.909]	[.897,.903]	[.899,.901]
.001	[.338,1.00]	[.754,.976]	[.861,.932]	[.888,.911]	[.897,.903]	[.899,.901]

10. Recapitulation

This book has been concerned to defend two main theses. The first concerns the nature of epistemic norms, and the second their content. The traditional view of epistemic norms had them functioning much like moral norms, guiding epistemic behavior in accordance with the intellectualist model and grounded in metaphysical facts about epistemic justification. We have argued that this view is wrong on both scores. It is logically impossible for epistemic norms to function in accordance with the intellectualist model, because that would lead to an infinite regress. And their grounding is in facts of human psychology rather than metaphysics. Epistemic norms are best viewed as descriptive of our procedural knowledge for how to cognize, and a theory of epistemic rationality is a competence theory of human epistemic cognition. The norms are still necessary truths, because they are partially constitutive of the concepts whose employment they govern, but they are not a priori. Instead, they are to be discovered by employing our built-in ability to detect when our behavior diverges from the norms comprising our procedural knowledge.

The preceding remarks characterize the nature of epistemic norms, but leave their content unspecified. The second main thesis of this book is that the epistemic norms comprising a competence theory of human epistemic cognition constitute a form of direct realism. Externalist theories of procedural justification can be eliminated on the grounds that internalizable norms must be internalist. Doxastic theories can be rejected on the grounds that they cannot accommodate perceptual knowledge. Specifically, (1) the beliefs that we acquire from perception are not self-justifying, so they must derive their justification from something other than their own existence, but (2) what justifies them is not other beliefs but rather the percept provided by perception.

Percepts license the adoption of beliefs about our immediate surroundings, and then various forms of inference license the adoption of beliefs that cannot be justified directly on the basis of perception. Such inference is typically defeasible, and we have sketched a theory of defeasible reasoning that is intended to capture its structure. Turning to the different kinds of defeasible inferences that are licensed by our epistemic norms, we have discussed temporal projection, certain kinds of causal reasoning, enumerative and statistical induction, and the use of the statistical syllogism to recover predictions about particular facts from statistical generalizations. An agent that can reason in accordance with these epistemic norms can perform many of the sophisticated epistemic tasks required of human beings, but not all. For example, we have not addressed knowledge of other minds, a priori knowledge, or the kind of means-end reasoning required in searching for plans for achieving goals. A complete procedural epistemology must eventually address all these topics at the low level we have explored in this chapter.

A novel, almost heretical, claim of this book is that epistemological theories cannot be adequately tested without implementing the proposed norms. The OSCAR project takes as its objective the construction of a general theory of rationality and the implementation of the theory in an AI system. Most of the epistemic norms described in this chapter have been implemented in OSCAR, and work is underway to implement the rest.

BIBLIOGRAPHY

Ackerman, Diana
1979   Proper names, propositional attitudes and non-descriptive con-
        notations. *Philosophical Studies* 35: 55-70.
1979a   Proper names, essences, and intuitive beliefs. *Theory and Decision*
        11: 5-26.
1980   Thinking about an object: Comments on Pollock. In *Midwest Studies
        in Philosophy*, vol. 5. Minneapolis: University of Minnesota Press,
        pp. 501-8.

Alston, William
1976   Two Types of Foundationalism. *The Journal of Philosophy* 73: 165-85.
1978   Meta-ethics and meta-epistemology. In *Values and Morals*, ed. A.
        I. Goldman and Jaegwon Kim. Dordrecht: D. Reidel, pp. 275-97.
1986   Internalism and Externalism in Epistemology. *Philosophical Topics*
        14: 179-221.
1988   An Internalist Externalism. *Synthese* 74: 265-83.

Anderson, John
1976   *Language, Memory, and Thought.* Hillsdale, NJ: Lawrence Erlbaum
        Associates.
1983   *The Architecture of Cognition.* Cambridge: Harvard University Press.
1995   *Cognitive Psychology and Its Implications.* New York: W. H. Freeman.

Annis, David
1969   A Note on Lehrer's Proof That Knowledge Entails Belief. *Analysis*
        29: 207-8.

Antony, Louise
1992   Quine as feminist: the radical import of naturalized epistemology.
        In *A Mind of One's Own: Feminist Essays in Reason and Objectivity*,
        ed. L. M. Antony and C. Witt. Boulder, CO: Westview, pp. 185-225.

Armstrong, David
1968   *A Materialist Theory of Mind.* London: Routledge & Kegan Paul.
1969   Does Knowledge Entail Belief? *Proceedings of the Aristotelian Society*
        70: 21-36.
1973   *Belief, Truth, and Knowledge.* London: Cambridge University Press.

Ayer, A. J.
1946   *Language, Truth, and Logic.* New York: Dover.

Bach, Kent
1982   *De re* belief and methodological solipsism. In *Thought and Object:*

Essays on Intentionality, ed. Andrew Woodfield. Oxford: Oxford University Press.

Baron-Cohen, Simon
1994 *Mindblindness: An essay on autism and theory of mind.* Cambridge: MIT Press.

Block, Ned
1986 Advertisement for a Semantics for Psychology. *Midwest Studies in Philosophy*, vol. 10: 615-678.

Bonjour, Laurence
1976 The coherence theory of empirical knowledge. *Philosophical Studies* 30: 281-312.
1978 Can empirical knowledge have a foundation? *American Philosophical Quarterly* 15: 1-14.
1980 Externalist Theories of Empirical Knowledge. In *Midwest Studies in Philosophy*, vol 5, ed. P. A. French, T. E. Uehling, and H. K. Wettstein. Minneapolis: University of Minnesota Press.
1985 *The Structure of Empirical Knowledge.* Cambridge: Harvard University Press.
1998 *In Defence of Pure Reason.* New York: Oxford University Press.

Carnap, Rudolf
1962 The aim of inductive logic. In *Logic, Methodology, and the Philosophy of Science*, ed. Ernest Nagel, Patrick Suppes, and Alfred Tarski. Stanford: Stanford University Press, pp. 303-18.
1967 *The Logical Structure of the World.* London: Routledge & Kegan Paul.
1971 Inductive logic and rational decisions. In *Studies in Inductive Logic and Probability*, ed. R. Carnap and R. C. Jeffrey. Berkeley: University of California Press.

Carroll, Lewis
1895 What the Tortoise Said to Achilles. *Mind* 4: 278-280.

Castaneda, H. N.
1966 He*: A study in the logic of self-consciousness. *Ratio* 8: 130-57.
1967 Indicators and quasi-indicators. *American Philosophical Quarterly* 4: 85-100.
1968 On the logic of attributions of self-knowledge to others. *Journal of Philosophy* 65: 439-56.

Cherniak, Christopher
1986 *Minimal Rationality.* Cambridge: MIT Press.

BIBLIOGRAPHY

Chisholm, Roderick
1957   *Perceiving*. Ithaca, NY: Cornell University Press.
1966   *Theory of Knowledge*. Englewood Cliffs, NJ: Prentice-Hall.
1977   *Theory of Knowledge*. 2nd ed. Englewood Cliffs, NJ: Prentice Hall.
1981   A version of foundationalism. *Midwest Studies in Philosophy*, vol.
       5, Minneapolis: University of Minnesota Press, pp. 543-64.
1981a *The First Person*. Minneapolis: University of Minnesota Press.

Clark, Michael
1963   Knowledge and grounds: A comment on Mr. Gettier's paper. *Analysis* 24: 46-48.

Clark, Romane
1973   Sensuous judgments. *Nous* 7: 45-56.

Clarke, Murray
1990   Epistemic Norms and Evolutionary Success. *Synthese* 85: 231-44.

Cohen, L. Jonathan
1981   Can human irrationality be experimentally demonstrated? *Behavioral and Brain Sciences* 4: 317-70.

Cohen, Stewart
1984   Justification and truth. *Philosophical Studies* 46: 279-96.

Creary, L. G., and C. J. Pollard
1985   A computational semantics for natural language. *Proceedings of the Association for Computational Linguistics, 1985*.

Cruz, Joseph
1999   *Epistemology in the scientific image*. Doctoral Thesis, University of Arizona.

Cummins, Robert
1989   *Meaning and Mental Representation*. Cambridge: MIT Press.
1996   *Representations, Targets, and Attitudes*. Cambridge: MIT Press.

Dalmiya, Vrinda, and Linda Alcoff
1993   Are "Old Wives' Tales" Justified? In *Feminist Epistemologies*, ed.
       L. Alcoff and E. Potter. New York: Routledge, pp. 217-44.

Dancy, Jonathan
1985   *Introduction to Contemporary Epistemology*. Oxford: Basil Blackwell.

Dennett, Daniel
1987   *The Intentional Stance*. Cambridge: MIT Press.
1995   Do Animals Have Beliefs? In *Comparative Approaches to Cognitive*

Science, ed. H. L. Roitblat and J. A. Meyer. Cambridge: MIT Press, pp. 111-118.

DePaul, Michael
1993 *Balance and Refinement: Beyond Coherence Methods of Moral Inquiry.* New York: Routledge.
1999 *Foundationalism.* Savage, MD: Rowman and Littlefield.

DePaul, Michael, and William Ramsey
1998 *Rethinking Intuition: The Psychology of Intuition and Its Role in Philosophical Inquiry.* Savage, MD: Rowman & Littlefield.

Donnellan, Keith
1972 Proper names and identifying descriptions. In *Semantics of Natural Language*, ed. Donald Davidson and Gilbert Harman. Dordrecht: D. Reidel.

Doyle, Jon
1979 A truth maintainance system. *Artificial Intelligence* 12: 231-72.
1982 Nonmonotonic logics. In *Handbook of Artificial Intelligence*, vol. 3, ed. Paul R. Cohen and Edward A. Feigenbaum. Los Altos, CA: William Kaufmann, pp. 114-19.

Dretske, Fred
1981 *Knowledge and the Flow of Information.* Cambridge: MIT Press.

Dummett, Michael
1975 What is a theory of meaning? In *Mind and Language*, ed. Samuel Guttenplan. Oxford: Oxford University Press.
1976 What is a theory of meaning? (II). In *Truth and Meaning*, ed. Gareth Evans and John McDowell. Oxford: Oxford University Press.

Eells, Ellory
1983 Objective probability theory theory. *Synthese* 57: 387-442.

Feldman, Richard
1985 Reliability and Justification. *The Monist* 68: 159-74.

Feldman, Richard, and Earl Connee
1985 Evidentialism. *Philosophical Studies* 48: 15-34.

Fetzer, James
1971 Dispositional probabilities. *Boston Studies in the Philosophy of Science*, vol. 8, pp. 473-82. Dordrecht: D. Reidel.
1977 Reichenbach, reference classes, and single case 'probabilities'. *Synthese* 34: 185-217.

1981   Scientific Knowledge. *Boston Studies in the Philosophy of Science*, vol. 69. Dordrecht: D. Reidel.

Field, Hartry
1977   Logic, Meaning, and Conceptual Role. *Journal of Philosophy* 69: 379-408.

Finetti, Bruno de
1937   La prevision: ses lois logiques, ses sources subjectives. *Annales de l'Institute Henri Poincare* 7: 1-68. Translated as Foresight: its logical laws, its subjective sources, in *Studies in Subjective Probability*, ed. Henry Kyburg and Howard Smokler, New York, 1964.

Firth, Roderick
1950   Radical empiricism and perceptual relativity. *The Philosophical Review* 59: 164-83, 319-31.
1978   Are epistemic concepts reducible to ethical concepts? In Goldman and Kim (1978).

Fodor, J. A.
1975   *The Language of Thought*. Cambridge: Harvard University Press.

Fodor, Jerry, and Ernest LePore
1992   *Holism: A Shopper's Guide*. Oxford: Basil Blackwell.

Foley, Richard
1992a   Being Knowingly Incoherent. *Nous* 26: 181-203.
1992b   The Epistemology of Belief and the Epistemology of Degrees of Belief. *American Philosophical Quarterly* 29: 526-34.
1994   In *Midwest Studies in Philosophy* vol. 19, ed. P. A. French, T. E. Uehling and H. K. Wettstein. Notre Dame: University of Notre Dame Press, pp. 243-60.

Fumerton, Richard
1994   Skepticism and Naturalistic Epistemology. In *Midwest Studies in Philosophy* vol. 19, ed. P. A. French, T. E. Uehling and H. K. Wettstein. Notre Dame: University of Notre Dame Press, pp. 321-40.
1995   *Metaepistemology and Skepticism*. Lanham, MD: Rowman and Littlefield.

Gelfond, Michael, and Vladimir Lifschitz
1993   Representing action and change by logic programs. *Journal of Logic Programming* 17: 301-22.

Gettier, Edmund
1963   Is justified true belief knowledge? *Analysis* 23: 121-23.

Giere, Ronald N.
1973 Objective single case probabilities and the foundations of statistics. In *Logic, Methodology, and Philosophy of Science*, ed. P. Suppes, et al. Amsterdam: North Holland.

1973a Review of Mellor's *The Matter of Chance*. *Ratio* 15: 149-55.
1976 A Laplacean formal semantics for single-case propensities. *Journal of Philosophical Logic* 5: 321-53.
1985 Philosophy of Science Naturalized. *Philosophy of Science* 52.
1990 *Explaining Science: A Cognitive Approach*. Chicago: University of Chicago Press.
1991 *Understanding Scientific Reasoning*. New York: Harcourt Brace.

Gigerenzer, Gerd
1991 How to make cognitive illusions disappear: Beyond heuristics and biases. *European Review of Social Psychology* 2: 83-115.
1996 On narrow norms and vague heuristics: A reply to Kahneman and Tversky. *Psychological Review* 103: 592-96.

Ginsberg, Matt
1987 *Readings in Nonmonotonic Reasoning*. Los Altos, CA: Morgan Kaufman.

Goldman, Alvin
1976 Discrimination and perceptual knowledge. *Journal of Philosophy* 73: 771-91.
1976a *A Theory of Human Action*. Princeton.
1979 What is justified belief? In *Justification and Knowledge*, ed. George Pappas. Dordrecht: D. Reidel.
1980 The internalist conception of justification. *Midwest Studies in Philosophy*, vol. 5. Minneapolis: University of Minnesota Press, pp. 27-52.
1985 The relation between epistemology and psychology. *Synthese* 64: 29-68.
1986 *Epistemology and Cognition*. Cambridge: Harvard University Press.
1988 Strong and weak justification. *Philosophical Perspectives 2: Epistemology*, ed. J. E. Tomberlin. Atascadero, CA: Ridgeview, pp. 51-69.
1992 Epistemic Folkways and Scientific Epistemology. *Liaisons*. Cambridge: MIT Press, pp. 155-75.
1994 Naturalistic Epistemology and Reliabilism. *Midwest Studies in Philosophy*, vol. 19, P. A. French, T. E. Uehling, and H. K. Wettstein. Notre Dame: University of Notre Dame Press, pp. 301-20.

Goldman, Alvin, and Jaegwon Kim (ed.)
1978 *Values and Morals*. Dordrecht: D. Reidel.

Goodman, Nelson
1951   *The Structure of Appearance.* Cambridge: Harvard University Press.
1955   *Fact, Fiction, and Forecast.* Cambridge: Harvard University Press.

Haack, Susan
1993   *Evidence and Inquiry.* Oxford: Basil Blackwell.

Hacking, Ian
1965   *Logic of Statistical Inference.* Cambridge: Cambridge University Press.

Hanks, Steve, and Drew McDermott
1986   Default reasoning, nonmonotonic logics, and the frame problem, AAAI-86.

Harman, Gilbert
1968   Knowledge, inference, and explanation. *American Philosophical Quarterly* 5: 164-73.
1970   Induction. In *Induction, Acceptance, and Rational Belief,* ed. Marshall Swain. Dordrecht: D. Reidel.
1973   *Thought.* Princeton: Princeton University Press.
1980   Reasoning and Explanatory Coherence. *American Philosophical Quarterly* 17: 151-82.
1981   Reasoning and evidence one does not possess. *Midwest Studies in Philosophy,* vol. 5, pp. 163-82. Minneapolis: University of Minnesota Press.
1984   Positive versus negative undermining in belief revision. *Nous* 18: 39-49.
1986   *Change in View.* Cambridge: MIT Press.

Heil, John
1983   Believing what one ought. *Journal of Philosophy* 80: 752-65.

Hempel, Carl
1962   Deductive-Nomological and Statistical Explanation. In *Minnesota Studies in the Philosophy of Science,* vol. 3, ed. Herbert Feigl and Grover Maxwell. Minneapolis: University of Minnesota Press.

Janlert, Lars-Erik
1987   Modeling change—the frame problem, in Z. Pylyshyn, ed. *The Robot's Dilemma,* MIT Press.

Jeffrey, Richard
1965   *The Logic of Decision.* New York: McGraw-Hill.
1970   Dracula Meets Wolfman: Acceptance vs. Partial Belief. In *Induction, Acceptance, and Rational Belief,* ed. Marshall Swain, pp. 157-85. Dordrecht: Reidel.

Kahneman, Daniel, Paul Slovic, and Amos Tversky
1982   *Judgment Under Uncertainty: Heuristics and Biases.* Cambridge: Cambridge University Press.

Kant, Immanuel
1958   *Critique of Pure Reason.* Translated by Norman Kemp Smith, London: Macmillan.

Kaplan, Mark
1985   It's not what you know that counts. *Journal of Philosophy* 92: 350-63.

Katz, Jerrold
1998   *Realistic Rationalism.* Cambridge: MIT Press.

Kim, Jaegwon
1988   What is "Naturalized Epistemology"? In *Philosophical Perspectives* 2, ed. J. E. Tomberlin. Atascadero, CA: Ridgeview Pub. Co., pp. 381-405.

Kintsch, W.
1974   *The Representation of Meaning in Memory.* Hillsdale, NJ: Lawrence Erlbaum Associates.

Kintsch, W., and J. M. Keenan
1973   Reading rate and retention as a function of the number of the propositions in the base structure of sentences. *Cognitive Psychology* 5: 257-74.

Kitcher, Philip
1992   The Naturalists Return. *The Philosophical Review* 101: 53-114.

Klein, Peter
1971   A proposed definition of propositional knowledge. *Journal of Philosophy* 68: 471-82.
1976   Knowledge, causality, and defeasibility. *Journal of Philosophy* 73: 792-812.
1979   Misleading "misleading defeaters". *Journal of Philosophy* 76: 382-86.
1980   Misleading evidence and the restoration of justification. *Philosophical Studies* 37: 81-89.

Kornblith, Hilary
1980   Beyond Foundationalism and the Coherence Theory. *The Journal of Philosophy* 72: 597-612.
1983   Justified belief and epistemically responsible action. *Philosophical Review* 92: 33-48.
1989   Introspection and Misdirection. *Australasian Journal of Philosophy* 67: 410-22.

248   BIBLIOGRAPHY

1993   *Inductive Inference and Its Natural Ground.* Cambridge: MIT Press.
1994a Naturalism: Both Metaphysical and Epistemological. In *Midwest Studies in Philosophy*, vol. 19, ed. P. A. French, T. E. Uehling, and H. K. Wettstein. Notre Dame: University of Notre Dame Press, pp. 39-52.
1994b *Naturalizing Epistemology, 2nd Edition.* Cambridge: MIT Press.

Koslowski, Barbara
1996   *Theory and Evidence: The Development of Scientific Reasoning.* Cambridge: MIT Press.

Kyburg, Henry, Jr.
1961   *Probability and the Logic of Rational Belief.* Middletown: Wesleyan University Press.
1970   Conjunctivitis. In *Induction, Acceptance, and Rational Belief*, ed. Marshall Swain. Dordrecht: D. Reidel.
1974   *The Logical Foundations of Statistical Inference.* Dordrecht: D. Reidel.

Lehrer, Keith
1965   Knowledge, truth, and evidence. *Analysis* 25: 168-75.
1968   Belief and knowledge. *Philosophical Review* 77: 491-99.
1971   How reasons give us knowledge, or the case of the gypsy lawyer. *Journal of Philosophy* 68: 311-13.
1974   *Knowledge.* Oxford: Oxford University Press.
1975   Reason and consistency. In *Analyses and Metaphysics*, ed. Keith Lehrer, pp. 57-74. Dordrecht: D. Reidel.
1979   The Gettier problem and the analysis of knowledge. In *Justification and Knowledge: New Studies in Epistemology*, ed. George Pappas. Dordrecht: D. Reidel, pp. 65-78.
1981   Self-profile. In *Profiles: Keith Lehrer*, ed. R. J. Bogdan. Dordrecht: D. Reidel.
1982   Knowledge, truth, and ontology. In *Language and Ontology*, proceedings of the 6th international Wittgenstein Symposium, ed. Werner Leinfellner. Vienna: Holder-Pichler-Tempsky.

1990a *Theory of Knowledge.* Boulder: Westview.
1990b *Metamind.* New York: Clarendon Press.

Lehrer, Keith, and Thomas Paxson
1969   Knowledge: Undefeated justified true belief. *Journal of Philosophy* 66: 225-37.

Levi, Isaac
1967   *Gambling with Truth: An Essay on Induction and the Aims of Science.* New York: Alfred A. Knopf.
1980   *The Enterprise of Knowledge.* Cambridge: MIT Press.

Lewis, C. I.
1946   *An Analysis of Knowledge and Valuation.* LaSalle: Open Court.
1956   *Mind and the World Order.* New York: Dover.

Lewis, David
1973   Causation. *Journal of Philosophy* 90: 556-67.
1979   Attitudes de dicto and de se. *Philosophical Review* 87: 513-43.
1980   A subjectivist's guide to objective chance. In *Ifs*, ed. W. L. Harper,
       R. Stalnaker, and G. Pearce, pp. 267-98. Dordrecht: D. Reidel.

Lin, Fangzhen, and Reiter, Raymond
1994   How to progress a database (and why) I. Logical foundations. In
       *Proceedings of the Fourth International Conference on Principles of
       Knowledge Representation (KR'94)*, pp. 425-36.
1995   How to progress a database II: The STRIPS connection. *IJCAI-95.*
       2001-2007.

Lipton, Peter
1993   *Inference to the Best Explanation.* New York: Routledge.

Lycan, William
1988   *Judgement and Justification.* New York: Cambridge University Press.

Maffie, James
1995   Towards an anthropology of epistemology. *The Philosophical Forum*
       36: 218-41.

Malcolm, Norman
1963   *Knowledge and Certainty.* Englewood Cliffs, NJ: Prentice-Hall, pp.
       229-30.

Margolis, Joseph
1973   Alternative strategies for the analysis of knowledge. *Canadian
       Journal of Philosophy* 2: 461-69.

McCarthy, John, and Hayes, Patrick
1969   Some philosophical problems from the standpoint of artificial
       intelligence. In B. Metzer & D. Michie (ed.), *Machine Intelligence 4.*
       Edinburgh: Edinburgh University Press.

McDermott, D., and Jon Doyle
1980   Non-monotonic logic I. *Artificial Intelligence* 13: 41-72.

Meiland, J.
1980   What ought we to believe? Or the ethics of belief revisited. *American
       Philosophical Quarterly* 17: 15-24.

Mellor, D. H.
1969 Chance. *Proceedings of the Aristotelian Society*, suppl. vol. 43, pp. 11-36.
1971 *The Matter of Chance.* Cambridge: Cambridge University Press.

Miller, George
1956 The magical number seven, plus or minus two: some limits on our capacity for processing information. *Psychological Review* 63: 81-97.

Millikan, Ruth
1984 Naturalist Reflections on Knowledge. *Pacific Philosophical Quarterly* 65(4): 315-34.

Moore, G. E.
1903 *Principia Ethica.* Cambridge: Cambridge University Press.
1959 Proof of an external world. In *Philosophical Papers*. London: Allen & Unwin.

Moser, Paul
1985 *Empirical Justification.* Dordrecht: D. Reidel.

Nelson, Lorraine
1993 Epistemological Communities. In *Feminist Epistemologies*, ed. L. Alcoff and E. Potter. New York: Routledge, pp. 121-59.

Neurath, Otto
1932 Protokollsatze. *Erkenntnis* 3: 204-14.

Newell, Allan
1972 A theoretical exploration of mechanisms for coding the stimulus. In *Coding Processes in Human Memory*, ed. A. W. Melton and E. Martin. Washington: Winston.
1973 Production systems: Models of control structures. In *Visual Information Processing*, ed. W. G. Chase. New York: Academic Press.
1980 Reasoning, problem solving, and decision processes: the problem space as a fundamental category. In *Attention and Performance VIII*, ed. R. Nickerson. Hillsdale, NJ: Lawrence Erlbaum Associates.

Newell, Allan, and H. A. Simon
1972 *Human Problem Solving.* Englewood Cliffs, NJ: Prentice-Hall.

Nisbett, Richard, and Lee Ross
1980 *Human Inference: Strategies and Shortcomings of Social Judgment.* Englewood Cliffs, NJ: Prentice-Hall.

Nozick, Robert
1981 *Philosophical Explanations.* Cambridge: Harvard University Press.

Pappas, George
1979   Basing Relations. *Justification and Knowledge*, ed., G. Pappas. Dordrecht: D. Reidel, pp. 1-23.

Perry, John
1977   Frege on demonstratives. *Philosophical Review* 86: 474-97.
1979   The problem of the essential indexical. *Nous* 13: 3-22.

Piatelli-Palmarini, Massimo
1994   *Inevitable illusions: How mistakes of reason rule our minds.* New York: Wiley and Co.

Plantinga, Alvin
1988   Positive Epistemic Status and Proper Function. *Philosophical Perspectives*, 2, ed., J. E. Tomberlin. Atascadero: Ridgeview, pp. 1-50.
1993a *Warrant: The Current Debate.* New York: Oxford University Press.
1993b *Warrant and Proper Function.* New York: Oxford University Press.

Pollock, John
1967   Criteria and our knowledge of the material world. *Philosophical Review* 76: 28-62.
1968   What is an epistemological problem? *American Philosophical Quarterly* 5: 183-90.
1970   The structure of epistemic justification. *American Philosophical Quarterly*, monograph series 4: 62-78.
1974   *Knowledge and Justification.* Princeton: Princeton University Press.
1979   A plethora of epistemological theories. In *Justification and Knowledge*, ed. George Pappas. Dordrecht: D. Reidel.
1980   Thinking about an object. *Midwest Studies in Philosophy* vol. 5, pp. 487-500. Minneapolis: University of Minnesota Press.
1981   Propositions and statements. *Pacific Philosophical Quarterly* 62: 3-16.
1982   *Language and Thought.* Princeton: Princeton University Press.
1983   A theory of direct inference. *Theory and Decision* 15: 29-96.
1983a Epistemology and probability. *Synthese* 55: 231-52.
1984   *The Foundations of Philosophical Semantics.* Princeton: Princeton University Press.
1984a Nomic probability. *Midwest Studies in Philosophy*, vol. 9, pp. 177-204. Minneapolis: University of Minnesota Press.
1984b A solution to the problem of induction. *Nous* 18: 423-62.
1984c Reliability and justified belief. *Canadian Journal of Philosophy* 14: 103-14.
1984d Foundations for direct inference. *Theory and Decision* 17: 221-56.
1986   A theory of moral reasoning. *Ethics* 96: 506-23.
1987   Epistemic norms. *Synthese* 71: 61-96.
1988   My brother, the machine. *Nous* 22: 173-212.
1990   *Nomic Probability and the Foundations of Induction.* New York: Oxford University Press.

1990   *How to Build a Person.* Cambridge: MIT Press.
1992   The theory of nomic probability. *Synthese* 90: 263-300.
1994   The projectibility constraint, in *Grue! The New Riddle of Induction,* ed. Douglas Stalker, Open Court, 135-52.
1994a   Justification and defeat. *Artificial Intelligence* 67: 377-408.
1995   *Cognitive Carpentry.* Cambridge: MIT Press.
1997   Reasoning about change and persistence: a solution to the frame problem. *Nous* 31: 143-169.
1998   Perceiving and reasoning about a changing world. *Computational Intelligence* 14: 498-562.
1998a   Procedural epistemology, in *The Digital Phoenix: How Computers Are ChangingPhilosophy,* ed. Terry Bynum and Jim Moor, Oxford: Basil Blackwell, 17-36.
1999   Procedural epistemology—at the interface of philosophy and AI". In *The Blackwell Guide to Epistemology,* ed. John Greco and Ernie Sosa. Oxford: Basil Blackwell.

Prakken, H., and G. A. W. Vreeswijk
2000   Logics for Defeasible Argumentation, to appear in *Handbook of Philosophical Logic, 2nd Edition,* ed. D. Gabbay. Dordrecht: Kluwer Academic Publishers.

Prichard, H. A.
1950   *Moral Obligation.* Oxford: Oxford University Press.

Putnam, Hilary
1960   Minds and machines. In *Dimensions of Mind,* ed. Sidney Hook. New York: New York University Press.
1975   The meaning of 'meaning'. In *Mind, Language and Reality,* ed. H Putnam. New York: Cambridge University Press.
1979   *Meaning and the Moral Sciences.* Cambridge: Cambridge University Press.
1984   *Reason, Truth, and History.* Cambridge: Cambridge University Press.

Quine, W. V. O.
1960   *Word and Object.* Cambridge: MIT Press.
1969   Epistemology Naturalized. In *Ontological Relativity and Other Essays.* New York: Columbia University Press.

Quine, W. V., and Joseph Ullian
1978   *The Web of Belief,* 2nd edition. New York: Random House.

Radford, Colin
1966   Knowledge—by examples. *Analysis* 27: 1-11.

Rapaport, William J.
1984   Belief representation and quasi-indicators. Technical Report 215, SUNY Buffalo Department of Computer Science.

Rapaport, William J., and Stuart C. Shapiro
1984 Quasi-indexical reference in propositional semantic networks. *Proceedings of the 10th International Conference on Computational Linguistics.* Morristown, NJ: Association for Computational Linguistics, 65-70.

Rawls, John
1971 *A Theory of Justice.* Cambridge: Harvard University Press.

Reichenbach, Hans
1949 *A Theory of Probability.* (Original German edition 1935.) Berkeley: University of California Press.

Reiter, Raymond
1978 On reasoning by default. *Theoretical Issues in Natural Language Processing* 2: 210-18.
1980 A logic for default reasoning. *Artificial Intelligence* 13: 81-132.

Rorty, Richard
1979 *Philosophy and the Mirror of Nature.* Princeton: Princeton University Press.

Ross, L., M. R. Lepper, and M. Hubbard
1975 Perseverance in self-perception and social perception: Biased attributional processes in the debriefing paradigm. *Journal of Personality and Social Psychology* 32: 880-92.

Ross, David
1930 *The Right and the Good.* Oxford: Oxford University Press.

Russell, Bertrand
1912 *Problems of Philosophy.* Oxford: Oxford University Press.

Ryle, Gilbert
1949 *The Concept of Mind.* Chicago: University of Chicago Press.

Salmon, Wesley
1966 *The Foundations of Statistical Inference.* Pittsburgh: University of Pittsburgh Press.

Savage, Leonard J.
1954 *The Foundations of Statistics.* New York: John Wiley.

Schiffer, Stephen
1981 Truth and the theory of content. In *Meaning and Understanding,* ed. H. Parret and J. Bouverese. Berlin: Walter de Gruyter.

Schlick, Moritz
1959   The foundations of knowledge. Reprinted in *Logical Positivism*, ed. A. J. Ayer, pp. 224-25. Glencoe, IL: Free Press.

Schmitt, Frederick
1984   Reliability, objectivity and the background of justification. *Australasian Journal of Philosophy* 62: 1-15.

Sellars, Wilfrid
1963   Empiricism and the philosophy of mind. Reprinted in *Science, Perception, and Reality*, New York: Humanities Press; London: Routledge & Kegan Paul.

Shoham, Yoav
1987   *Reasoning about Change*. Cambridge, MA: MIT Press.

Shope, Robert K.
1983   *The Analysis of Knowing*. Princeton: Princeton University Press.

Skyrms, Brian
1967   The explication of 'X knows that p'. *Journal of Philosophy* 64: 373-89.

Smith, David Woodruff
1984   Content and context of perception. *Synthese* 61: 61-87.
1986   The ins and outs of perception. *Philosophical Studies* 49: 187-212.

Sosa, Ernest
1964   The analysis of 'knowledge that p'. *Analysis* 25: 1-8.
1974   How do you know? *American Philosophical Quarterly* 11: 113-22.
1980   Epistemic presupposition. In *Justification and Knowledge: New Studies in Epistemology*, ed. George Pappas. Dordrecht: D. Reidel.
1981   The raft and the pyramid: coherence versus foundations in the theory of knowledge. *Midwest Studies in Philosophy*, vol. 5, pp. 3-26. Minneapolis: University of Minnesota Press.

Squires, Robert
1969   Memory unchained. *The Philosophical Review* 77: 178-97.

Stevenson, Charles L.
1944   *Ethics and Language*. New Haven: Yale University Press.

Strawson, P. F.
1952   *An Introduction to Logical Theory*. London: Methuen.

Stich, Stephen
1990   *Fragmentation of Reason*. Cambridge, MA: MIT Press.

Stein, Edward
1995 *Without Good Reason.* New York: Oxford University Press

Stroud, Barry
1984 *The Significance of Philosophical Skepticism.* New York: Clarendon Press.

Suppes, Patrick
1973 New foundations for objective probability: Axioms for propensities. In *Logic, Methodology, and Philosophy of Science*, vol. 5, ed. Suppes et al., 515-29. Amsterdam: North Holland.

Swain, Marshall
1981 *Reasons and Knowledge.* Ithaca, NY: Cornell University Press.

Taylor, James
1990 Epistemic Justification and Psychological Realism. *Synthese* 85: 199-230.

Touretzky, David
1984 Implicit orderings of defaults in inheritance systems. *Proceedings of AAAI-84.*

Thagard, Paul
1982 From the descriptive to the normative in psychology and logic. *Philosophy of Science* 49: 24-42.

Tye, Michael
1984 The adverbial approach to visual experience. *The Philosophical Review* 93: 195-226.

Unger, Peter
1967 Experience and factual knowledge. *Journal of Philosophy* 64: 152-73.

Van Cleve, James
1979 Foundationalism, epistemic principles, and the Cartesian Circle. *Philosophical Review*, 88: 55-91.

Wason, P.C., and P. N. Johnson-Laird
1972 Psychology of Reasoning: Structure and Content. London: B. T. Batsford.

Williams, Michael
1977 *Groundless Belief.* New Haven: Yale University Press.
1991 *Unnatural Doubts: Epistemological Realism and the Basis of Scepticism.* Oxford: Basil Blackwell.

Winograd, Terry
1980   Extended inference modes in reasoning by computer systems. *Artificial Intelligence* 13: 5-26.

Wittgenstein, Ludwig
1953   *Philosophical Investigations*, 3rd edition.   Translated by G. E. M. Anscombe. New York: Macmillan.

INDEX
(bold text indicates main entry)

ABOUT THE AUTHORS

John Pollock is professor of philosophy and research professor of cognitive science at the University of Arizona. He works primarily in epistemology, philosophical logic, and artificial intelligence. He directs the OSCAR project, whose purpose is the construction of a general theory of rational cognition and its implementation in an artificial agent. Recent books include *Cognitive Carpentry, How to Build a Person,* and *Nomic Probability and the Foundations of Induction.* John is also an inveterate mountain biker and outdoorsman. Web address: http://www.u.arizona.edu/~pollock/.

Joseph Cruz earned the Ph.D. in philosophy and cognitive science from the University of Arizona in 1999, and is currently assistant professor of philosophy and cognitive science in Hampshire College's School of Cognitive Science. In philosophy, he works primarily in the philosophy of psychology and epistemology. His interests in empirical aspects of cognitive science focus on cognitive development and the psychology of reasoning. Joe is currently working on the nature of psychological explanation and the foundations of scientific psychology. He has published in *Behavioral and Brain Sciences* and *Mind and Language.* Joe shares John's love for the out-of-doors, even when they are hopelessly lost in the backcountry. Web address: http://hampshire.edu/~jlcCCS/.